The Wind & The Wizard

VOLUME 1

*"The only hope, or else despair
Lies in the choice of pyre or pyre —
To be redeemed from fire by fire."*
— Eliot

To John Bradshaw,
Champion of the wonder child,
with appreciation,

Richard Roberts

THE WIND AND THE WIZARD, © 1990 by Richard Roberts. All rights reserved in all countries. No part of this book may be reproduced or transmitted in any form or by any means, electronic, or mechanical, including photocopying, recording, or by any information storage and retrieval system, without written permission from the publisher.

For information write Vernal Equinox Press, P.O. Box 581, San Anselmo CA. 94960.

Books
by Richard Roberts

TAROT AND YOU (1970)

TAROT REVELATIONS (1979)
with Joseph Campbell

TALES FOR JUNG FOLK (1983)

FROM EDEN TO EROS (1984)

THE ORIGINAL TAROT AND YOU (1987)

THE WIND AND THE WIZARD (1990)

The Wind & The Wizard

VOLUME 1

Richard Roberts

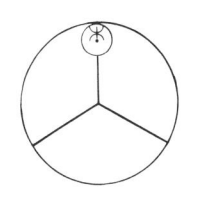

VERNAL EQUINOX PRESS,
P.O. BOX 581,
SAN ANSELMO CA. 94960.

*This book is dedicated to
Joseph Campbell (1904-1987)
Stephen Hawking
Barbrey and Bongo
and RJ*

Illustration by Elizabeth Gill
Design by Judith Alwood

Word Processing by Glenna Goulet

Special thanks to Robert Baum, Jr. for permission to use
passages and illustrations from L. Frank Baum's
The Lost Princess of Oz (1917)

Library of Congress Cataloging-in-Publication Data.
Richard Roberts
    The Wind and the Wizard
Richard Roberts; illustrations by Elizabeth Gill
ISBN 0-942380-07-X for volume one

I. Fiction   88-40518

CONTENTS

VOLUME 1

INTRODUCTION
VI - VII

BOOK ONE
EARTH
10 - 65

BOOK TWO
WATER
66 - 125

BOOK THREE
AIR
126 -175

BOOK FOUR
WOOD
176 - 239

INTRODUCTION

When Heinrich Zimmer died prematurely in 1943 he left uncompleted a monumental body of work, later edited and completed by the mythologist Joseph Campbell. One of these volumes, *The King and the Corpse*, contains the story of a king who had for ten years received daily a fruit offering from a holy man in the garb of a beggar. Passing the fruit along to his treasurer who stood behind the throne, the king gave the gift no further thought, and the beggar withdrew each time into the crowd.

One day in the tenth year, a monkey came bounding into the hall and leaped onto the arm of the throne, becoming a king of a kind for a moment. The beggar had just presented his daily offering, and the king playfully gave the fruit to the monkey, who bit into it. Suddenly a valuable gem, hidden in the husk, dropped out and rolled across the floor.

The eyes of everyone in the throne room grew wide with wonder. The king turned to the treasurer at his shoulder. "What has become of the rest?" he asked. The treasurer had been discarding the fruit by tossing them over his shoulder through an open window into a distant alcove of the treasury. Rushing there they found a mass of dried or rotting husks of fruit and, amidst the decade's debris, a great heap of priceless jewels. In the works of Jung, reference is made to the alchemists who sought gold or the "treasure hard to attain," the Self, or spiritual center of the psyche, far within the many layers of the husk of personality. At the same time the alchemists tell of the contempt of the common man for this jewel, which is often discarded on a midden heap, as in *The King and the Corpse*.

The rest of the story does not concern us here; it is the "accidental" happening of the opening of the fruit which does. In April of 1987 I was having dinner in Honolulu with Joseph Campbell and his wife, when I told him my idea for a novel in which a young boy's monkey opens ("cracks") a book, and instantly he and the monkey are in the book as characters, their adventure interwoven with the book's characters.

Joseph Campbell paused to reflect for a moment, no doubt reminded of *The King and the Corpse*, and said "That's a very good idea, Dick!"

This was the last time I was to see my friend; he died in October of that year. But I have the fondest memory of that moment and dinner at the Outrigger Club at the edge of Waikiki Beach, catamarans bending to a gentle breeze, an Hawaiian trio serenading us with "Lovely Hanalei" (where I was to vacation in Kauai), while the sun set symbolically and oh! so beautifully over an ocean of sea and sky.

There is some pain in the memory too, for I realize the experience cannot be repeated, and that our soul-sharing of more than twenty years has now ended. During that time Joseph Campbell became my inspirational mentor. Now the Self would have to be a mentor unto myself.

Had Joseph Campbell not thought this book to be such a good idea, I'm not sure I would have completed it. It has taken three years in the preparation and writing, and required two volumes. Little did I know that our collaboration in writing *Tarot Revelations* was to be a kind of alchemical Great Work preparing me for the writing of this book. Realization of Selfhood at the end of the individuation process is attained by a sacred "marriage" of conscious and unconscious minds; or expressed another way, this marriage harmonizes left and right brains.

The idea for this book and the writing thereof came about as a result of this fascinating process. Readers interested in such matters may read about the process in the appendix at the end of the book, wherein I also discuss the book's structure. As far as I know, no author has ever created characters that actually go into another book and interact with that book's characters.

"To what purpose?" one may ask. This book is the adventure of a seven year-old "hero with a thousand faces," destined to bring the world new models of space/time as a theoretical physicist in his adult life, one well prepared for his role by the spiritual instruction he receives in each adventure. But adventure is the key to spiritual discovery, as Joseph Campbell has taught us, and while each book is enlightening it is also adventurous, entertaining, and humorous, in keeping with the original spirit of the classic books into which my hero ventures. Further, it is hoped that this book will stimulate public interest in reading once again these classics. These books represent worlds which I once loved and into which I thought I could never venture again. But I was wrong. Imagination opened these portals once more, and like Bertie and Bongo I became a traveller in space/time.

Lastly, at the end of the chapters I have placed a list of definitions of words which may not be known to younger children. As we know, children in England received a better education 50-100 years ago in the English language than they do now in the U.S. and Britain. The child need only turn to the end of each chapter in order to read the meaning of a word he or she may not know; thus the book becomes comprehensible to readers down to about age seven.

Early stimulation of the faculty of imagination may be the most important factor in influencing a child to become an artist, writer, musician, or an innovative scientific thinker. I know it was true in my life; therefore, I have resurrected the marvelous language, characters, and morality of the very best of children's literature, which have survived so well because ultimately they endear themselves to one's adult mind, looking back nostalgically to "the land of lost content ... the happy highways where I went/And cannot come again."

(A. E. Housman:*A Shropshire Lad*)

"Time present and time past
Are both perhaps present in time future,
And time future contained in time past."
- T. S. Eliot

The author and his sister

PROLOGUE

It had flown many miles across the English Channel from its launching point on the coast of France. Its purpose and that of the rest of its dread kind was to kill as many English men, women, and children as possible without distinction between soldier, sailor, or civilian. When its moment of revenge approached, its roar ceased, and imperceptibly at first, and then faster and faster it began to fall on the English countryside. This one found the home of a British scientist and his wife, a writer of children's stories, killing them instantly. Outside, their only child Bertram had been sleeping in his pram at the bottom of the garden. Awakened by the horrendous blast that leveled his house and took from him his mother and father, he wailed the centuries-old cry of an orphan.

BOOK ONE

EARTH

Heard the hero's call
And made his own way

BOOK ONE

Five years had passed, and Bertie was now living with an aunt in Kensington. Because she was an old maid and knew nothing about children, she treated him as one might a puppy, keeping him in a huge library in one room of her mansion. Each day three meals and a high tea of jam and cakes were brought to the library by servants. Once a week he was led out for a bath, whether he needed it or not.

He seldom saw his aunt, nor heard anything about the outside world, and so he read, there being really nothing else to do in a library. He was precocious certainly, blooming early under the nurturing influence of his scientific father and his creative mother. She had taught him to read at the age of two, and how a dictionary and all of its words were simply the alphabet in sequence. With the dictionary as his tool, he had gone on to conquer new worlds each day, taking new words for himself as easily as a soldier might take prisoners.

Having no one else with whom to discuss the books he read each day, for as I have said, there was nothing else to do in a library, he talked to his only companion Bongo, a stuffed monkey who had been with him as early as he could remember. Sometimes when a book was particularly provocative, Bertie would pause in his reading to Bongo, who sat on his knee in a huge leather chair near a bay window that looked out to Kensington Gardens and the distant, beckoning statue of Peter Pan, which he had seen one day a very long time ago when a young servant girl had taken him to the park for "air" when his aunt had gone away for a day. When Bertie's aunt learned of it, she sacked the girl, and that ended all that he was to know of the world except that which he read in books and became colored by his imagination.

He wanted to be a physicist like his father, who had named him after the famous Bertrand Russell, so Bertie often read scientific books, which led to many discussions with Bongo.

"Bongie," Bertie would say, "do you think the universe began with a Big Bang, or has it always existed?"

He would think long on this, and then go on reading until another idea triggered his wondering.

"Bongie, if there are Black Holes in the universe we could be swallowed by one as likely as not, don't you think?"

This time Bongo nodded his head in agreement.

"I wonder what it would be like," said Bertie.

Because Bertie had never known any other children, he thought that every child in the world was learning the same way; hence he did not feel deprived or unhappy. Children read libraries, he thought, and then came out and were physicists or bus conductors, or whatever.

And so his life went on the same way each day, until one day a very strange thing happened. Bongo opened a book, whether accidentally or on purpose we shall never know, but suddenly........!

Nature's stately procession had marched to a different drummer each season along the rippling river's bank. First had come a corps of Panpipes, flutes, and shrill fifes heralding spring's dainty approach, when reluctantly she had revealed naught but the hem of her verdant gown. Summer's full brass gave way to September's languorous woodwinds, yielding in turn to October's chilling strings, which gave ground to November's cello deeps, culminating in a final gloomy drum that sounded December's death knell and sent the wood and riverbank into a long January slumber, ne'er to waken until hearing once again, when the year's symphony had played full circle, Panpipes borne on an errant March wind blown from the cerulean Mediterranean. And what were the brooks along the river bank babbling? "Intruders on the road!"

Bongo and Bertie stood hand in hand along a roadside somewhere they knew not and heard in the distance a honking of horns coming from a cacophonous cloud of dust that moved rapidly down the road towards them.

"I wonder if it's a tornado or a Black Hole?" said Bertie.

There was a squeal of brakes and the dust cloud moved on down the road, leaving behind the most curious sight Bongo and Bertie had ever seen, an old-time automobile with behind the wheel a frog or toad wearing racing goggles. Next to him sat a mole, and in the back a rather large badger and rat. By itself the spectacle of four animals taking an outing in a motor car was curious enough, but the most curious thing of all was that each animal was outfitted quite dapperly in a full suit of clothes.

The toad lifted his goggles and set them on top of his head in order to get a better look at Bongo and Bertie. Meanwhile, the mole began to wail. "O, don't stop, don't stop, dear Toady. The whatzit is quite naked."

"Decorum suggests that we avert our eyes," said the badger, placing a paw over his.

"A scandal! A genuine scandal!" said the toad excitedly, standing up.

"Oh, sit down, Toad," said the rat sensibly. "You're the one who is causing the scandal."

Bertie had quite enough of their remarks and spoke out. "Why on earth would you expect a stuffed monkey to be wearing clothes?" he said.

"What's it called?" asked the rat. "A moon-key?"

"What's it stuffed with?" the toad questioned.

"Oh, I don't know!" Bertie replied, losing his patience. "Can't you help us?!"

"No need to get huffy," said the mole, starting to sniffle.

"My dear young sir," said the badger, "we were not aware that you were seeking aid, else we would have obliged you at the earliest opportunity."

"Here, here," said the toad.

"Tell us then," inquired the rat, "what is the matter."

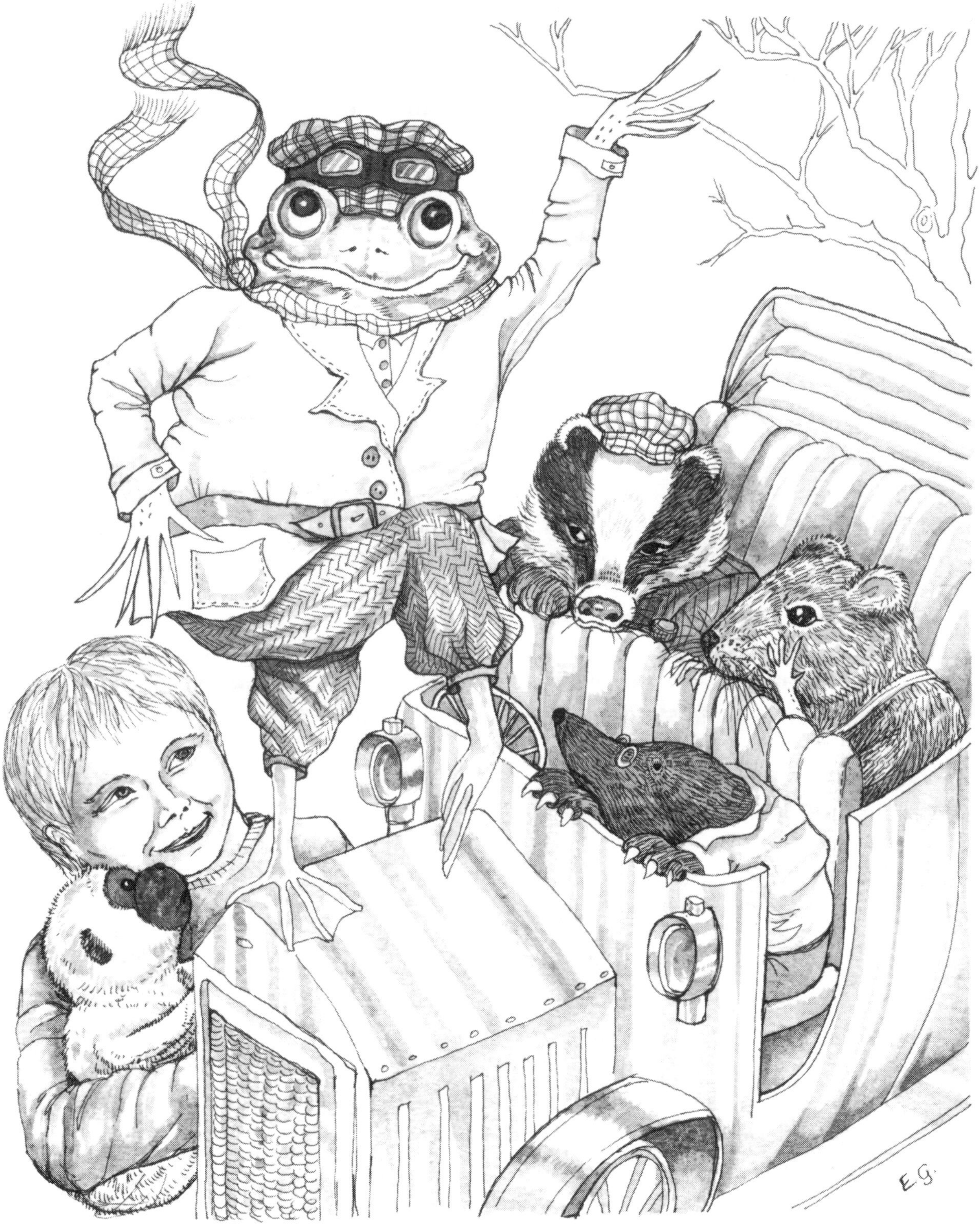
E.G.

"Well," said Bertie, taking a deep breath, "I was sitting in my aunt's library in London, England, in the year nineteen hundred and fifty, when all of a sudden my monkey here, Bongo, started to open a book and then—Bam!—We found ourselves deposited in a heap alongside a river..."

"The River," corrected the rat.

"...Just down the road from here," Bertie continued.

"Can you remember the name of the book your Bongo had opened?" said the badger.

"Why, yes, it was called *The Wind in the Willows*."

Toad jumped up on the hood and did a delightful little jig. "I tell you chaps, they've landed in our story!"

"Oh, my, oh, my," said the mole, starting to sniffle again.

"But how it that possible?" said Bertie.

"I'm sure I don't know," said the rat, "but now that you're here, you'd better make the most of it. Allow me to make the introductions. To my left is Badger. Occupying the passenger seat is Mole. Our driver, risking life and limb for his latest thrill is the notorious Toad of Toad Hall."

Toad made a long, sweeping bow.

"Oh, how you do go on, Raty," said Mole.

"I myself am Water Rat, or just Rat for short, an author of some river lays and ditties for solo voice, not unknown completely in these parts."

"But where are these parts?" beseeched Bertie, totally perplexed at finding himself a character in someone else's book.

"Well, you haven't left England, if it's any consolation," said Badger.

"Bang on, Badger," said Toad, "we're British! I'm sure enough of that!"

"Oh, yes, indeed," said Mole, "and Church of England, too."

Rat burst into song.

"'And did those feet, in ancient times,
Walk on England's green and pleasant land.'"

"Yes," said Badger, "it's a very pleasant place here, if its any consolation to you."

A dreamy look came over Mole's face. "Oh, yes, indeed. How very, very pleasant it is. And there is nothing quite so pleasant as a picnic. All those tins and jars spread on the grass with the River gently gliding by." And he closed his eyes and dreamily began to recite a litany of all the things he had had on his very first picnic with Rat on that day when tired of his spring cleaning, he climbed up the long tunnel from his underground home, and entered the bright world of the woodland, oh, such a long time ago.

"Coldchickencoldtonguecoldhamcoldbeefpickledgherkinssaladfrenchrollscressandwidgepottedmeatgingerbeerlemonadesodawater."

He was about to start again when Toad interrupted. "Poop-poop!" he said, making the sound of a car's horn. "Picnics are fine enough, Moley, but nothing can match the thrill of a motor car tearing along the open road."

15

And then he was off. His eyeballs began to spin like pinwheels in his head, and his body became rigid, and while sitting on the hood, he assumed the posture of the driver of a car. His arms and elbows swayed back and forth as if piloting a car through hairpin curves, and his legs—though rigid—began to bounce up and down on the hood of the car.

"Oh, he's off again," said Badger, disgustedly.

But Toad's display was upstaged almost at once by the voice of Water Rat, who proclaimed at the top of his lungs to all of the world, "Believe me, my young friend, there is nothing—absolutely nothing—half so much worth doing as simply messing about in boats. simply messing," he went on in a dreamy singsong, "messing—about—in—boats—messing— about—in—boats—messing—about—in—boats...."

Badger shook his head. "I'm afraid you'll have to forgive my friends. Each has an all-consuming passion in life and sometimes they do get out of hand."

"Well, we should like to do all of those things with you," said Bertie. "Go on a picnic, ride in your motorcar, and mess about in boats, but...."

"Messing—messing about in boats," corrected Rat.

"All well and good, but with whom would you be messing about?" said Bertie. "You haven't asked us our names!"

"Well, then, introduce yourselves," said Badger. "The naked creature you called 'Bongo.'"

"Bongo, he is," said Bertie, "and he's a monkey."

"I'm sure there are no others like him—not even in the Wild Wood," said Mole, coming out of his reverie.

"He's a cute little fellow," said Badger. "Could he be a cousin to humans?"

"A very distant cousin," replied Bertie.

"Come on, old chap," said Toad, coming to his senses. "Tell us who you are."

"My name is Bertram, but I like to be called Bertie."

"A very decent name, I'm sure," said Rat, "but in our story we're all called by our animal names, so we shall have to call you 'boy,' if you don't mind, and your friend is a...."

"Monkey," said Bertie. "But I do mind. I'm used to being called Bertie. A servant boy might be called 'boy' because no one would care what his name was, but I am Bertie," and he stamped his foot on the running board for emphasis.

"If I'm not being too presumptuous, chaps," said Mole, "just for the time that these two are in our story, that is, for the duration of their visit...."

"Stop beating about the bush, Mole," said Toad impatiently.

"Yes, do get to the point, dear Mole," added Rat.

"What I mean is, why couldn't we take on Christian names, since we're all Church of England."

"A capital idea!" said Toad.

"Then our friends here would be free to use their given names."

"Only thing to do to make them feel at home," said Badger wisely.

"Let me see," said Toad, "I rather fancy Timothy Toad, although you could all call me Tim."

"Melvin Mole has a nice subterranean ring to it, don't you think, chaps?"

"Roderick Badger," intoned Badger, drawing out all the vowels. "Now there's a name to be reckoned with."

"No, on second thought," said Toad, "I like the name Trevelyan Toad...Sir Trevelyan Toad."

"What about a name for you, Raty?" said Mole.

"Well, I don't know...Mustn't rush into it...bears considering, or one could be stuck with it."

"Sir Trevelyan Toad might not fit on the coat of arms," interrupted Toad, changing his mind again.

"Toad!" said Rat, very much annoyed, "you change names as often as your famous fixations. Last week it was the gypsy caravan, 'the song of the open road, the dusty highway, the green heaths, the perilous blind hedgerows, the rolling downs! Here today and up and off to somewhere else tomorrow! Travel, change, interest, excitement! The whole world before you, and a horizon that's always changing!' Those were your very words."

"But that was last week!" cried Toad.

"It may have been last week, but my bones still ache from the crash we took when the cart overturned in a ditch."

"My head still hurts," added Mole, taking the Rat's part.

"And yet you expect your friends," continued Rat, "to go along with you on whatever hare-brained scheme flits in and out of that careless cavity you call a brain."

"Well put indeed, old chap," said Badger.

"Really, dear Toad, it is a bit much," chimed in Mole.

"But don't you see, I'm done with carts forever. If it hadn't been for you, dear friends, I never would have seen that swan, that sunbeam, that thunderbolt, that glorious motorcar that upset the cart and turned my life over too! No more indecision, chaps, it's Motorcars Forever!"

"If I may interrupt your story," said Bertie, "it may not be too late to save him."

"Your words have an ominous ring," said Badger. "Save whom and from what?"

"Why, to save Toad from the police and imprisonment."

"Oh, I knew it," wailed Mole, and he began to cry uncontrollably.

"The police get me," scoffed Toad, "not on your life." And he began, whilst standing on the car's hood, a kind of rapid-fire ballet in which he mimed drawing pistols from his waist and shooting at the encircling arm of the law. And all the while as he danced and spun about, contemptuous invective for his attackers poured from his mouth. "Take that copper! You'll never take me alive. Come on, you dirty rats."

"That does it!" cried Rat, "I'm going back to The River."

"I must say, that was an ill-chosen choice of words, Toad, even for one as insensitive as yourself to the feelings of others," said Mole.

"I won't stand for racial remarks," said Rat. "Contrary to being dirty, we Water-Rats are among the most hygienic of any animal."

"Quite right, Mole," seconded Badger.

"Pigeon Toad," spoke Bertie.

"What?" they said, and all turned to look at him.

"Pigeon would be a good name for Toad, because he's all puffed up and full of himself like one of those strutting, unbearably overbearing birds."

It took but a moment for Bertie's words to take effect, and then *each* animal except Toad (who was momentarily taken aback at the boy's cheekiness) began to chuckle, which *they* tried unsuccessfully to stifle, seeing the rising color in Toad, whose skin began to change ever so suddenly from green to pink and then to a pulsating vermillion. Soon the chuckles grew to full-sized guffaws, presently maturing to belly laughs of wrenching proportions, culminating in gigantic howls, which carried on the wind all the way to the Wildwood. The wind itself could not resist its own cyclone of laughter, which came as it was passing one of the Wildwood's most venerable and, therefore, unsteady of ancient oaks, blowing it to the ground as a cacophony of crows and jackdaws that had been roosting in its branches rose into the air reverberating the raucous chorus begun so long ago with a stifled chuckle.

The scene alongside the road where Toad's motorcar had stopped for Bongo and Bertie was now one of total chaos, for in succession Badger, Mole, and finally Rat had tumbled out of the car and onto the ground. Their laughter, too great for the car to contain, attracted other animals within earshot and they peered from holes, perched on logs, sat on hind legs, or paused in flight amazed. From a distant vantage point it appeared that Toad, standing rigid at the very front of the hood was an ornament like a winged Mercury, or some such embellishment which an owner might choose to adorn his fine vehicle. But the oddest thing about the striking figure of the rigid Toad was the steam which emitted from his ears in ever more rapid spurts. In the end, he could stand no more and commandeered the wheel, nearly tearing it from the steering column as he wrenched the car away from the site of his humiliation and hurtled down the road with the speed of a lightning bolt, despite the "Come back!" cries of his companions left hanging in the air like departing dust.

Our companions had been walking a weary way, all the while reviling the infamous Toad, who had left them on the road, (no more to be seen), when the sun began to set, tinging the sky with the faintest of pink, a color to be seen in those parts only in the earliest spring primroses. But the season was not spring, but mid-summer, and so the hour was not far from being midnight.

"Oh, bother," said Rat, "now we shall never make it to The River until morning, and my legs cannot carry me another mile."

"Why not stay with me, Ratty?" said Mole. "It's often enough I've enjoyed your hospitality of late, and my house is just across that distant field there."

The weary travelers paused to look in the direction Mole indicated, and it was agreed that they would all, except for the intrepid Badger (whose house lay half way between Rat's and Mole's) stay what remained of the short night in Mole's snug harbor beneath the ground.

For Mole it was a most welcome chance to return once again to the domicile where resided such sweet memories, and from which he had departed so suddenly in the midst of his spring-cleaning earlier that very year. Then spring was moving its subtle tendrils in the air above, and below in the earth around him he felt the roots of giant oaks stirring as their fingers sought greater strength to sustain them in the long growing season ahead.

For the sensible Mole, a sudden impulse like this had never before come upon him. He had finished all the sweeping and dusting and had commenced applying his annual coat of whitewash to the walls inside when the spirit of divine discontent overwhelmed him. Flinging his whitewash brush to the floor, he said, in rapid succession, "Bother," "O blow," and "Hang spring-cleaning," and bolted out the door without even waiting to put on his coat.

Something up above called to him as compelling as a magistrate's summons, not to be denied, nor even reflected upon, and he made a mole-line for the steep tunnel which served as his own grand carriage drive. Scratching and scraping with his nose and paws his snout popped out at last in a meadow's milky-warm grass. The sun stroked his fur, and he exclaimed out loud, "How wonderful, a fire that doesn't have to be stoked "All the hard work of pushing up into the light had heated his brow, and now as if on cue an obliging breeze cooled it for him. His hearing, dulled by so many months snug in the earth's depths, now was rewarded by a serenade of birds, singing their own joy at the return of spring.

And on that day, he fell in love, as Rat had himself fallen so many years ago with the same fair creature. Never in his life had he seen a river before—This sleek, sinuous, full-bodied animal, chasing and chuckling, gripping things with a gurgle and leaving them with a laugh, to fling itself on fresh playmates that shook themselves free, and were caught and held again. All was a-shake and a-shiver—glints and gleams and sparkles, rustle and swirl, chatter and bubble. The Mole was bewitched, entranced, fascinated. By the side of the river he trotted as one trots, when very small, by the side of a man who holds one spellbound by exciting stories; and when, tired at last, he sat on the bank, while the river still chattered on to him, a babbling procession of the best stories in the world, sent from the heart of the earth to be told at last to the insatiable sea.

And so it was, on that very day, that Mole met Rat; messed about in a boat for the first time; and became Rat's inseparable companion, residing with him at his bijou waterside residence. But just before falling asleep at night, two tiny tears often formed in his eyes at the thought of his home, forlorn without him.

Now he welcomed the chance to show Rat his humble house, and particularly Bertie, who was telling Mole quite excitedly that he'd never been in a real home underground, except the bomb shelter during the war, and that didn't really count as an honest-to-goodness house.

"Mind you now, Master Bertie, you'll have to crawl on your hands and knees down the tunnel before we get to the front door."

"Oh, I won't mind," said Bertie. "I used to crawl as a baby a long, long time ago, but I still remember."

"What shall we do about Bongo?" said Mole, pausing at the entry tunnel.

"He could ride on my back," said Bertie.

"No, no," said Mole, "that just won't do. You'll have to get very low to make it yourself. I know! Ratty, would you mind carrying Bongo?"

"Well," said Rat humorously, "I've already got 'messing-about-in-boats' for a habit. I guess it wouldn't hurt to have another monkey on my back."

"Dear Ratty, how you do go on," chuckled Mole.

Once inside the pitch-black tunnel Bertie began to have second thoughts. The earth smell was strong and the passage was close, but every time he stopped, Rat's nose prodded him from behind, so he pressed on, albeit mentally noting to bawl out Bongo when they were alone for what he had gotten them into by opening *The Wind in the Willows*. After what seemed ages, the passage ended and they could all stand erect. Mole struck a match and reached down a lantern hung from a nail on the wall before them. By its light, Bertie could see that they were at Mole's front door, for the name "Mole End" was painted in Gothic letters across the door.

Immensely relieved at no longer having to crawl down the claustrophic passage, which aroused heretofore unconscious memories of his own birth, Bertie made a little joke, "Is this the End, then?" he said.

But as soon as he said it, he wanted to bite his tongue, for Mole had turned his face against the door and was blubbering like a baby.

Bertie began apologizing, but Mole cut him short.

"My dear chap, it's not your remark that makes me weep, but the sight of these objects so dear to me from which I have been so long absent. And now, seeing them," and he turned round, "brings back all of those happy memories."

And as he spoke, Bertie and Rat looked in the direction in which Mole was gazing tearfully.

They saw that they were in a sort of forecourt. At the side of the door was an elaborately carved wooden bench, so that travelers could rest in the interval between pulling the bell pull to summon Mole and the moment of his appearance. Resting its handle against the bench was a garden roller, for uncharacteristically Mole could not stand seeing his ground all humped up. On a wall near the door, amid hanging ferns which made a kind of grotto of the place, hung a picture of a rather severe Queen Victoria. Down one side of this forecourt ran a miniature golf course—three holes only—with what once had been carefully manicured greens, with little tubes and tunnels for the balls to follow, and now in the three months since his absence were so overgrown as to be impassable even to the best hit balls, which were actually marbles fallen from the torn pockets of some boy passing on the main road.

"Oh, dear, oh, dear," said Mole, viewing his overgrown greens. "You must come in. I stand here sniffling while you all have come such a long way."

And so saying, he reached into his pocket and quickly unlocked the door for them. Hurrying ahead with the lantern, he led the way into a combination parlor-kitchen with shelves and cupboards everywhere. Seeing the thick layer of dust lying on everything, Mole, who was a Virgo, started sniffling again, but Rat saved the day by darting everywhere, pulling out drawers and inspecting rooms and cupboards, all the while giving a running commentary. "What a capital little house this is. So compact. So well thought out. Did you design it yourself? Say no more, I thought so."

And so Mole was thoroughly heartened, and he set about fashioning for his guests the finest of midnight suppers. Just before they were about to turn in for the night, with a bed for each one except Bongo, who always slept with Bertie, they heard a scurrying in the forecourt outside the door. It was Mole's friends the field mice. They had seen his light and dropped by to invite him to the Midsummer Eve festival that very next evening at Stonehenge, but Mole, without mentioning Toad or his many adventures, had to disappoint them by saying he would be returning tomorrow to the waterfront chalet of his friend Rat.

When all was quiet, and the last pawfall of the mice had died in the tunnel up to the world, Bertie whispered to Bongo, who was curled up with his head under Bertie's chin, "You got us into this. How do you propose to get us out?"

Bongo said nothing. From the next room came the contented snoring of Mole.

That night Bertie dreamed that he and Bongo had taken a taxi from their aunt's home to their old house in Hampstead where mother and father were waiting for them. On the way, Bertie tried to make conversation with the driver, calling him cabbie.

"Oh, I'm not a cabbie, governor," said the red-faced man, "I'm a chimney-sweep."

"Then what business do you have picking up people in cabs?" questioned Bertie somewhat angrily.

"You and me, gov, are going to take a little trip in a motor car."

And as they watched from the rearview mirror his features gradually became atavistic, turning monkey-like at first, and then very gradually amphibian, so that Bongo and Bertie confronted in the mirror none other than the furious face of the infamous toad, who drove the taxi at an ever accelerating pace. Clinging in fear to each other, Bongo and Bertie witnessed in rapid succession on each block as they passed, a pageant of English kings, back to the first in history, receding even unto the mythical Arthur. Then, as the taxi accelerated like a rocketship, all the houses on the blocks in London town began falling down, like a line of dominoes collapsing, and Bertie knew he was going back, back, back in time. Suddenly the cab became a pram and Bertie a baby again. The rearview mirror expanded, and he saw in it his father's and mother's faces, their arms held out to him. He was trying to come to them as the pram approached the speed of light and they receded before him just out of reach, into blackness, into infinity.

"Ouch," said Bertie. "Darn." The first thing he had done upon awakening was to pinch himself when he saw that he was evidently still inside "Mole End" far below the ground. "That's how you tell whether or not you're dreaming," he told Bongo, still snuggled under his chin. Then he gave Bongo a pinch for good measure. "When that hurts, you'll know you're alive."

"How can I be inside a book?" he thought. "There is nothing in the known laws of physics that permits this. I'll bet Bertrand Russell would like to know about this."

Now, from the next room, came the voice of Rat reading to Mole. "And so he goes on to say, 'Space is infinite not by grand extension, but by re-entry.'"

As he stood in the doorway listening, suddenly all of his dream came back to him.

"Where did you get that book?" demanded Bertie, astounded at the sight of Rat reading to Mole.

"Oh, we animal folk are not as primitive as you humans make us out to be," replied Rat good-naturedly.

"With all due respect, Mr. Rat," said Bertie contritely, "I don't remember you or Mole reading a book in *The Wind and the Willows*. It was one of my favorite stories—my mother read it to me at least twice."

Then turning around he was suddenly confronted by a huge bookcase of books reaching from floor to ceiling. "I could have sworn those weren't there when I came in last night."

"Oh, you musn't swear, dear boy," said Rat. "Moley doesn't like it—at least not in the house."

"You're quite right, of course, Master Bertie, I never did have any books in that story. Too busy tidying up to have any time for reading."

"But Rat and I have been up quite early having a very serious discussion about you and your rather peculiar predicament."

"Damned peculiar, I'd call it," said Rat, and then began to apologize to Mole for his language.

"No need; it is damned peculiar. But you yourself, dear Ratty, are known in these parts to have composed a ditty or two. Tell me, how often have you heard tell of a person becoming a character in another's story?"

"Never. No, positively, it has never happened before. And I'll warrant there isn't a single such story in all the British Library."

"There, you see," pronounced Mole, "it is damned peculiar if it has never happened before. You see, dear Master Bertie, we moles are accustomed to getting to the bottom of things from digging in the earth, ergo it follows that the crucial question to which we must aid you in finding an answer is, 'To what purpose do you appear in our story? And as a corollary, why does your monkey accompany you?'"

"Moley," interrupted Rat, "why this sudden metaphysical inclination? Never before were you interested in such matters, indeed, you never told me you even read a book."

"Quite right, Ratty. You do see what's going on, don't you? I sensed it last night when we first walked in. You've never been here before, so you don't realize what's happened. Let me see if I can get to the bottom of this. Master Bertie, precisely what does golf mean to you?"

"Well, nothing really...only...."

"Yes, yes, do go on."

"Well, my father was going to teach me golf before he...got...killed...by the flying bomb."

"Flying bombs! Good heavens," exclaimed Rat. "I hope that doesn't come during Victoria's reign."

"Oh, no," said Bertie, "not until the 1940's."

"Now Ratty, you often asked me about my hobby and I told you how much I enjoyed playing skittles."

"And you said the skittles alley was in the forecourt at your front door! " exclaimed Rat, suddenly very excited. "Well, where is it, Moley?"

"I ask you. Or perhaps we should ask our young friend here."

"...who prefers golf," shouted Rat, jumping off his chair.

"Precisely my point," concluded Mole. "He's changing our story."

"And the books must now be here," added Rat, "because Bertie feels at home with them, or he hopes to find within them answers as to his purpose in being here in our story."

"Now there is yet another possibility," discoursed Mole. "Bertie may not be real after all, but himself a character in a book in which we are also appearing, but again to what purpose I do not know."

"May I?" said Bertie, holding up his hand as if in a classroom. "If that line of reasoning is extended, then the author of the book in which I am appearing, may also be a character in a book, the author of which book may or may not himself be real."

"Now you've really set my head to spinning," exclaimed Mole, throwing up his hands. "Oh, what has happened to the simple animal pleasures while I played the philosopher. Come up to the table and have some breakfast, for goodness sakes. There's rashers of bacon and I'll make hot chocolate for all of us, and Rat can compose a ditty about changing space and time, and afterwards, a round of golf, or perhaps something else if by that time you have a greater preference," concluded Mole, winking at Bertie from behind his spectacles.

When the breakfast dishes had been cleared away, Mole brought out delicate clay pipes for himself and Rat, and when the fires of their sweet tobacco had been kindled, they sat back from the table to determine a course of action.

"We shall help you in finding your purpose, if you will help us with the Toad," said Rat.

"Fair enough," replied Bertie. "I have one advantage, you don't have; I know where your story is going."

"Unless of course you change it some more," added Mole.

My it was pleasant to be in Mole End with a cheery fire burning in the grate, thought Mole to himself. Let Toad fend for himself. He'd rather be here or picnicking on The River rather than setting Toad's world aright. Then Bertie uttered the awful words that shattered the peace of Mole End.

"He's in jail."

"Who?!" said Rat and Mole simultaneously.

"Toad. For stealing a car."

"But why would he steal a car when he already has one?" queried Rat.

"Because you and Badger wouldn't let him use it. You lock him up in Toad Hall, but he tricks you, Rat, by feigning illness, and asking you to fetch a doctor. Then he climbs out a window, steals a car, and lands in jail. But he tricks everyone, and comes swaggering back to boast about his exploits, not contrite at all, and a bigger braggart than before."

"I'm not sure I want to be involved in such a sordid story," said Mole, starting to sniffle.

"Stiff upper lip there, Moley," said Rat, "after all Toad is our friend, and a friend in need is a friend indeed."

At mid-day our friends knocked at Badger's door deep in the Wildwood. Surely sensible Badger would know what to do about Toad. They found Badger in none too good a mood, for they had interrupted his noon nap. Although fully dressed, his down-at-heels carpet slippers betrayed his protestations that they had not awakened him.

"Old friends are always welcome here at any time of the day or night," he said, none too convincingly.

As they followed Rat, who led the way behind Badger down the long, gloomy passageway that opened into a sort of central hall, Mole whispered to Bertie about the night when lost in the snow they had chanced upon Badger's brass door plate, saving—if not their lives—a night of certain unpleasantness in the keen wind and whirling snow.

Long wanting to make the acquaintance of Badger, Mole had unwisely ventured out alone into the Wild Wood, leaving Rat in his armchair by the fire alternately dozing and working over rhymes that wouldn't fit. It was a cold, still afternoon with a hard steely sky overhead, when Mole slipped out of Rat's warm parlor into the open air. The country lay bare and entirely leafless around him, and he thought that he had never seen so far and intimately into the inside of things as on that winter day when Nature was deep in her annual slumber and seemed to have kicked off her clothes.

Copses, dells, quarries and all hidden places, which had been mysterious mines for exploration in leafy summer, now exposed themselves and their secrets pathetically, and seemed to ask to overlook their shabby poverty for a while, till they could riot in rich masquerade as before, and trick and entice him with the old deceptions. It was pitiful in a way, and yet cheering—even exhilarating. He was glad that he liked the country undecorated, hard, and stripped of its finery. He did not want the warm clover and the lay of seeding grasses; the screens of quickset, the billowy drapery of beech and elm seemed best away; and with great cheerfulness of spirit he pushed on towards the Wild Wood, which lay before him low and threatening, like a black reef in some southern sea.

There was nothing to alarm him at first entry. Twigs crackled under his feet, logs tripped him, fungi on stumps resembled caricatures, and startled him for the moment by their likeness to something familiar and far away; but that was all fun and exciting. It led him on, and he penetrated to where the light was less, and trees crouched nearer and nearer, and holes made ugly mouths at him on either side.

Everything was very still now. The dusk advanced on him steadily, rapidly, gathering in behind and before; and the light seemed to be draining away like floodwater.

Then the faces began.

It was over his shoulder, and indistinctly, that he first thought he saw a face: a little evil wedge-shaped face, looking out at him from a hole. When he turned and confronted it, the thing had vanished.

He quickened his pace, telling himself cheerfully not to begin imagining things, or there would be simply no end to it. He passed another hole, and another; and then—yes!—no!—yes! certainly a little narrow face, with hard eyes, had flashed up for an instant from a hole, and was gone. He hesitated—braced himself up for an effort and strode on. Then suddenly, as if it had been so all the time, every hole, far and near, and there were hundreds of them, seemed to possess its face, coming and going rapidly, all fixing on him glances of malice and hatred; all hard-eyed and evil and sharp.

If he could only get away from the holes in the banks, he thought, there would be no more faces. He swung off the path and plunged into the untrodden places of the wood.

Then the whistling began.

Then faint and shrill it was, and far behind him, when first he heard it; but somehow it made him hurry forward. Then, still very faint and shrill, it sounded far ahead of him, and made him hesitate and want to go back.

Then the pattering began.

He thought it was only falling leaves at first, so slight and delicate was the sound of it. Then as it grew it took a regular rhythm, and he knew it for nothing else but the pat-pat-pat of little feet, still a very long way off. Was it in front or behind? It grew and multiplied, till from every quarter as he listened anxiously, leaning this way and that, it seemed to be closing in on him.

The pattering increased till it sounded like sudden hail on the dry-leaf carpet spread around him. The whole wood seemed running now, running hard, hunting, chasing, closing in round something or—somebody? In panic, he began to run too, aimlessly, he knew not whither. He ran up against things, he fell over things and into things, he darted under things and dodged round things. At last he took refuge in the deep dark hollow of an old beech tree, which offered shelter, concealment—perhaps even safety, but who could tell? Anyhow, he was too tired to run any further and could only snuggle down into the dry leaves which had drifted into the hollow and hope he was safe for the time. And as he lay there panting and trembling, and listened to the whistlings and patterings outside, he knew it at last, in all its fullness, that dread thing which other little dwellers in field and hedgerow had encountered here, and known as their darkest moment—that thing, which the Rat had vainly tried to shield him from—the Terror of the Wild Wood!

Fortunately for Mole, Rat had noticed that Mole's cap was missing from its accustomed peg, and his galoshes were also gone. Following Mole's tracks into the Wild Wood, he had patiently been searching for an hour or more, all the time calling out, "Moley, Moley, Moley! Where are you? It's me—it's old Rat!"

At last to his joy he heard a little answering cry, and crept into the hollow at the foot of an old beech tree, and there he found the Mole, exhausted and still trembling. "O Rat!" he cried, "I've been so frightened, you can't think."

"And then, Master Bertie," said Mole concluding his story, "we both set out together to find Badger's home, and do you know, because the snow had piled up so high, I tripped over Badger's door-scraper and cut my shin, and if I hadn't done that we never would have found his front door, and that small brass nameplate on which his name was engraved. It seems almost comical now, but I can tell you truthfully, Master Bertie, I was a bit frightened.

"Well, I should have been, too," replied Bertie.

"You would have been frightened too? Well, that does make me feel a bit better."

Badger did not take kindly, when they were seated in his kitchen, to Bertie's suggestion that in a future chapter Toad would outwit Badger and Rat, although Badger allowed as how indeed he had been thinking about locking Toad in to prevent his running amuck on the highway. Badger asked Bertie to tell again the story of Toad's folly as he remembered hearing it when read to by his mother. Then, when Bertie had finished he pushed back his great chair at the head of the table and stood up.

"It seems to me the hardest part of the story is getting them there stoats and weasels out of Toad Hall. Why don't we go there now, and head them off at the pass, so-to-speak," and he chuckled to himself.

"A capital idea, Badger," said Rat.

"This time I'll really learn 'em," cried Badger, grasping a stout cudgel from next to the fireplace.

"Perhaps that won't be necessary this time," said Bertie, with a twinkle in his eye. "I think the author has forgotten something."

The animals looked quizzically at Bertie, but he said no more.

Now memory as we know becomes imperfect with time, and it was that Bertie's memory of our story, as read to him by his mother some years before, was somewhat fuzzy about the edges. Perhaps he was falling asleep at this or that point in the narrative, or perhaps he was asleep altogether, and his mother had not noticed and kept on reading. Mothers get sleepy too, you know.

At the time she was reading *The Wind in the Willows* to Bertie, she was working on one of her own stories, *Helpful Mouse*, which became her most famous book, although she never had more than a modest success with her career as a writer of children's stories. She had met her husband while they were both students at university. Despite the weighty concepts he expounded on physics and the creation of the universe, she found him boyishly attractive, with his close-cropped hair and constant pipe, which he often forgot to light. They fell in love at once, and were married within a year. As one of the university's rising stars, he was given a graduate teaching fellowship that enabled them to just get by, with occasional help from their middle-class parents.

It was while thinking of her husband's clutter lying about the room that the thought of Helpful Mouse came to her as suddenly as a mouse might appear at its door in the wainscot. Picking up after her husband was an endless task.

She could practically follow him from room to room, and by the time he returned to the same room which she had first tidied, he would have it cluttered again. Books came out of their shelves, opened, and were left on a table, chair, or floor whenever a new idea hurtled into his head and made him drop the old line of thought wherever it fell. Clothes dropped anywhere, and often disappeared—particularly soxs. To give him a key was to have it disappear.

And so, one day she thought, "If only there could be a Helpful Mouse for him—no, for everyone—to follow us around, or watch from the wainscot, and then retrieve the lost or forgotten things, placing them back in their proper places so we can find them when we look for them again.

And then when Bertie came along, the idea came back to her again, and new ideas flew in. Beyond tidying up a child's room while he slept, Helpful Mouse became a true friend. When a child thought the world did not understand, or was being unfair, Helpful Mouse explained, and made things seem better all around. And when nothing went right and tears came, Helpful Mouse curled up and slept comfortably under the child's chin, and during the night, Something Helpful happened, for the next day was always so much brighter than the day before.

By the time Bertie was one, she had the book finished, and even had painted the watercolor illustrations. It sold five thousand copies the first year, and later was even reprinted in America.

There were many characters like Helpful Mouse who were to come to her in her future, to be written down, and dressed up, and given a life of their own. But before they could come, the flying bomb came instead and all those characters had to wait, somewhere out in the universe, for their creator to come again.

When Bertie's mother read to him, he often interrupted to ask the meaning of a word he did not know, then he would ask her to say it once again very slowly, and then he would pronounce it and file it away in his memory. All the characters in *The Wind in the Willows* were alive in his memory when he went into the book; yet in another sense they were alive somewhere else waiting for him. The universe has many mysteries, the greatest of which may be the imagination.

That very night in Badger's house, Bertie had told our friends the future of their story, which of course they did not know since it was in their *future*, while for Bertie it was in the past, that time when his mother had read it to him. By telling it to them, he changed it somewhat, and as they slept and dreamed that night, they changed their story again as their dreaming selves created new realities that their waking bodies would act upon.

Now as they sleep, telling and retelling their fabled adventure to themselves, this is the story for *you* alone of Toad's adventure, without the fuzziness of Bertie's memory.

When Toad found himself imprisoned in a damp and smelly dungeon, and knew that all the grim darkness of a medieval fortress lay between him and the outer world of sunshine and well-paved high roads where he had lately been so happy, disporting himself as if he had bought up every road in England, he flung himself at full length on the floor, and shed bitter tears, and abandoned himself to dark despair. "This is the end of everything," he said, "at least it is the end of the career of Toad, the rich and hospitable Toad, the Toad so free and careless and debonair! How can I hope to be ever set at large again," he said, "who have been imprisoned so justly for stealing so handsome a motorcar in such an audacious manner, and for such lurid and imaginative cheek, bestowed upon such a number of fat, red-faced policemen!" Here his sobs choked him. "Stupid animal that I was," he said, "now I must languish in this dungeon, till people who were proud to say they knew me have forgotten the very name of Toad! O wise old Badger!" he said, "O clever, intelligent Rat and sensible Mole! What sound judgments, what a knowledge of men and matters you possess! O unhappy and forsaken Toad!"

Now the jailer had a daughter, a pleasant wench and good-hearted. She was particularly fond of animals, and kept several piebald mice and a restless revolving squirrel. This kind-hearted girl, pitying the misery of Toad, said to her father one day, "Father! I can't bear to see the poor beast so unhappy, and getting so thin! You let me have the managing of him. You know how fond of animals I am. I'll make him eat from my hand, and sit up and do all sorts of things."

Her father replied that she could do what she liked with him. He was tired of Toad, and his meanness. So that day she went on her errand of mercy, and knocked at the door of Toad's cell.

"Now cheer up, Toad," she said coaxingly, on entering, "and sit up and dry your eyes and be a sensible animal. And do try to eat a bit of dinner. See, I've brought you some of mine, hot from the oven!"

The penetrating smell of cabbage reached the nose of Toad as he lay prostrate in his misery on the floor, and gave him the idea for a moment that perhaps life was not such a blank and desperate thing as he had imagined. But still he wailed, and kicked with his legs, and refused to be comforted. So the wise girl retired for the time, but, of course, a good deal of the smell of the hot cabbage remained behind, as it will do, and Toad, between his sobs, sniffed and reflected, and gradually began to think new and inspiring thoughts: of chivalry, and poetry, and deeds still to be done; of broad meadows, and cattle browsing in them, raked by sun and wind; of kitchen gardens, and straight herb borders, and warm snapdragons beset by bees; and of the comforting clink of dishes set down on the table at Toad Hall, and the friendly faces of Badger, Mole, and Rat glowing in the firelight. Yes, if he ever got out he would not let them down again.

And when the jailer's daughter returned to fetch his tray, she had a plan that would free him. Next evening the girl ushered her aunt into Toad's cell, bearing his week's laundry pinned up in a towel. In return for ten pounds of hard cash, Toad received a cotton print gown, an apron, a shawl, and a bonnet; the only stipulation being that the old lady be bound and gagged so as to give the appearance that Toad had overpowered her. Next came the difficult part—getting Toad into her dress. Too many years of grand banquets had taken their toll on the Toad's waistline, and there was many an anxious moment before the jailer's daughter was able to "hook-and-eye" him into the gown.

Then he was off. The washerwoman's squat figure in its familiar print dress seemed a passport for every barred door and grim gateway. Once outside he walked quickly towards the lights of the town, where he made straight for the station, consulted a timetable, and found that a train, bound more or less in the direction of his home, was due to leave in half an hour. His spirits rising rapidly, he went off to the booking office to buy his ticket. He gave the name of the station nearest the village of which Toad Hall was the principal feature, and reached into where his pocket should have been in order to pay the fare. To his horror he realized that all his money was in the waistcoat left behind him in the cell.

In his misery he made one desperate effort to carry the thing off. His voice a blend of the Squire and the College Don, he said, "Look here, old chap, I find I've left my purse behind. Just give me the ticket, will you, and I'll send the money on tomorrow with a little honorarium for your kind consideration. I am, of course, well known in these parts."

The clerk stared at him in his washerwoman attire and then laughed. "I should think you were pretty well known in these parts, if you've tried that old badger game before. Here, stand away from the window now; you're obstructing the other passengers."

Baffled and full of despair, he wandered blindly down the platform.

"Hullo, mother," said the train engineer, "what's the trouble?"

Grasping his opportunity, Toad poured out a tale designed to soften the hardest heart.

"Well, I'll tell you what I'll do," said the good engineer, and in no time a bargain was struck whereby in exchange for a ride to his destination, Toad was to wash the engineer's shirts when he got home, and then send them along to him.

Once on board and with every mile taking him further and further away from the dreaded prison, Toad began to skip up and down and sing snatches of songs in praise of himself and his great cleverness, to the astonishment of the train engineer, who had come across washerwomen before, but never one like this.

They had covered many miles when Toad saw the engineer climb onto the coals and peer over the top of the train. "That's strange," he said, "there's another train following us."

Toad's heart sank, and he knew there was but one course to take. Leaping from the train, he rolled down an embankment and came to rest against the side of a tree. Shortly thereafter, the pursuing train came into view. It was an engine only, crowded with the oddest assortment of people. Prison guards, waving halberds; policemen in their helmets, waving truncheons, and shabbily dressed men in bowler hats, unmistakably plain clothes detectives, waving revolvers; and all of them waving and shouting the same thing, "Stop, stop, stop!"

Poor Toad, even as the train passed by and he was safe once more, he could not stop shuddering and shaking even though it was a summer night. He dared not leave the shelter of the trees, so he struck into the wood, with the idea of leaving the railway as far as possible behind him. After so many days behind walls, he found the wood strange and unfriendly. Nightjars, sounding their mechanical rattle, made him think that the wood was full of searching wardens closing in on him. An owl, swooping noiselessly towards him, brushed his shoulder with its wing, making him jump with the horrid certainty that it was a hand; then flitted off, moth-like, hooting its low ho! ho! ho!At last, cold and hungry, and all tired out, he sought the shelter of a hollow tree, where with branches and leaves he made as comfortable a bed as he could and slept.

The next morning, having dawned warm and felicitous, our friends lolled about on the river bank, trying to determine the best course to take. Badger was once more for action. Sensibly he explained that if the secret passage to Toad Hall had worked once before, surprising the stoats and weasels in their lair, it should work once again. Badger was certainly shocked to learn that Bertie knew about it, since Toad's father had revealed it to Badger in the strictest confidence. But Bertie's knowledge of it did much to enhance his credibility with them.

"No, I don't think it will be necessary to use the secret passage this time," said Bertie.

"Well, we done it before according to you," said Badger, "but we have only your word on it, which is good enough, mind you, but what I mean to say is, well, darn it all, if we can't recall having done it, then we h'aint done it yet, and it would be great fun to do it."

"You see, Bertie," added Rat, "I think what Badger is driving at is that you are talking about an event which is in your past, since your mother read to you the part about our recapturing Toad Hall, via that secret passage, whereas for us this event is in *our* future, so we don't know that it will happen unless we experience it."

"I understand," Bertie replied, "but one can never be sure that an event will happen exactly as before. After all, my appearance in your story has already changed it somewhat, and you might have been lucky before—"

"There you go again," interrupted Rat, "referring to our future as if it was already past."

"Oh, dear," sniffed Mole, "time can be so confusing. What does your watch say, Master Bertie, are you in the past, present, or future?"

Bertie looked at his watch and was startled. "Why, it's stopped!" he said.

"Then you are *out* of time," observed Rat.

"Everyone runs out of time," Badger remarked.

"Eureka!" declared Bertie. "That must be my purpose then, to get back into time."

Rat watched his remark, hanging in the air, and then carefully added. "Perhaps so, Master Bertie, but until you discover why you came into this story—taking you out of time—it would be presumptuous, don't you think, to say that your purpose is to go back into the world of time. After all, if a door has opened out of time for you, you don't just go back through that same door without finding out why it opened for you in the first place."

"Here, here, Raty," Mole chimed in.

"Of course, you're quite right," Bertie responded. "I do know there must be some reason why I'm here with you. Bongo probably knows better than I, because he opened the book, but he won't say."

"All this talk leaves me cold," Badger blurted out. "I'm an animal of action, always have been, always will be, and Rat's a natural born phil-losopher. Doesn't make him any better than me, nor any worse, but I say we do something. Anything."

"All right," Bertie agreed, "suppose we start looking for Toad today."

And well they should, for as we know Toad had already escaped, and having spent the night in the hollow of a tree awakened that very morning to the same bright sunlight. The exceeding coldness of his toes made him dream that he was at home in bed with the Tudor window open and his toes peeping out from under the blankets, where they began to grumble about his lack of consideration for them. Finally, they could stand the cold no longer and decided to flee down the stairs to the kitchen where they could toast themselves before the warmth of the kitchen fire. Toad leaped out of bed to follow them, but without his toes he couldn't run very well, and kept lurching from side to side, knocking into the furniture and banging his shins in the bargain. When he reached the top of the stairs he was totally out of control and began a long tumble that carried him to the bottom. He jumped out again, beseeching his toes to be reasonable and return to him, but still they ran on, and it was then that he awoke.

There was just room in the tree hollow for him to sit upright, but even with his knees bent, his toes hung outside the sheltering roots of the tree, and a cold early-morning wind had nipped at his toes. Recalling their unpleasant departure in his dream, he reached down with his paws to count them, and heard in his head his mother's voice reciting a nursery rhyme: "This little toady went to market; this little toady stayed home." Satisfied that they were all there, he looked around for the familiar walls of his prison home; then, with a leap of his heart, remembered everything—his escape, his flight, his pursuit; remembered, first and best of all that he was free!

Free! The word and thought alone were worth losing all of his toes if need be. He was warm from head to toe as he thought of the jolly world awaiting his triumphal re-entry, ready to serve and play up to him, as it had been in days of old before misfortune befell him. No matter, he would turn his incarceration to his advantage by writing a monograph—"The Art of Prison Breaking." Once more himself, he combed the dry leaves and twigs out of his hair where some early swallows had begun a nest, and started out for Toad Hall, cold but confident, hungry but hopeful.

He had all the world to himself, that early summer morning. The dewy woodland as he walked it, was solitary and still; the green farm fields that succeeded the trees like an unravelling panorama were his own to do as he liked with; and the road itself, when he reached it, in that early morning loneliness that was everywhere, seemed like a stray dog, to be looking anxiously for someone to walk with it.

Toad, however, was looking for something that could talk, and tell him clearly which way he ought to go. It is all very well, when you have a light heart, and a clear conscience, and money in your pocket, and nobody scouring the country for you to drag you off to prison again. But practical Toad cared very much indeed where the road was bound, and could have kicked it for its passive silence when every minute was of importance to him.

The reserved rustic road was presently joined by a shy little brother in the form of a canal, which took its hand and ambled along by its side in perfect confidence, but with the same tongue-tied uncommunicative attitude towards strangers. "Bother them!" said Toad to himself. "But, anyhow, one thing's clear. They must both be coming *from* somewhere, and going *to* somewhere. You can't ignore that, Toad, my boy!" So he marched on patiently by the water's edge.

Round a bend in the canal came plodding a solitary horse, stooping forward as if in anxious thought. From rope traces attached to his collar stretched a long line, taut, but dipping with his stride, the further part of it dripping pearly drops. Toad let the horse pass, and stood waiting for what the fates were sending him.

With a pleasant swirl of quiet water at its blunt bow the barge slid up alongside of him, its gaily painted gunwale level with the towing path, its sole occupant a big stout woman wearing a linen sunbonnet, one brawny arm laid along the tiller.

"A nice morning, ma'am!" she remarked to Toad, as she drew up level with him.

Forgetting for the moment his washerwomen attire, and the depressing fact that he had no money, Toad was about to charter the barge for a triumphal return to Toad Hall, arriving with all the splendor of Henry VIII's royal barge returning to the palace at Richmond. But he caught himself in time, and replied politely, "I dare say it is, ma'am," as he walked along the towpath abreast of her. "I dare say it is a nice morning to them that's not in sore trouble like what I am," Toad continued. "Here's my married daughter, she sends off to me posthaste to come to her at once; so off I comes, fearing the worst. I've left my business—I'm in the washing and laundering line—and left my young children to look after themselves, and a more troublesome set of young imps doesn't exist, ma'am; and I've lost all my money, and lost my way."

"Where might your married daughter be living, ma'am?" asked the barge woman.

"She lives near the river, ma'am," replied Toad. "Close to a fine house called Toad Hall."

"Toad Hall? Why, I'm going that way meself," said the barge woman, who steered the barge close to the bank, and Toad, with many humble and grateful acknowledgements, stepped lightly on board and sat down with great satisfaction. "Toad's luck again!" thought he. "I always come out on top."

"So you're in the washing business?" inquired the barge woman as they glided along.

"Finest business in the whole country," said Toad airily. "I simply dote on it. Never so happy as when I've got both arms in the washtub."

"What a bit of luck meeting you," said the barge woman. "*I* like washing, too, but my husband who by rights should be doing the steering has gone off with the dog to see if he can pick up a rabbit for dinner. Now there's a heap of my things that you'll find in a corner of the cabin. If you'll just take one or two of the most necessary sort—I won't venture to describe my unmentionables to a lady like you—and just put them through the washtub it will be a real help to me and you can earn your passage, so-to-speak."

Well, despite Toad's protestations she would not take no for an answer, and soon Toad found himself up to his elbows in her washtub, while a pile as high as his head of her unmentionables and other outer garments loomed ominously nearby.

But clever Toad, always seeking to grasp victory from the jaws of defeat, thought of a plan. Undoubtedly a description of his washerwoman's disguise had gone out from the prison guards, the station master, and all present there, so the jig would be up soon, very likely unless, he thought, he were to change his disguise. With this in mind, he began stuffing pieces of the barge woman's apparel down his bodice. Once shut of her when he had reached the vicinity of Toad Hall, he could change in the twinkling of an eye and no one would be the wiser. And so as he worked, he laughed to himself how he was getting the better of the barge woman, when all at once she appeared suddenly behind his back.

"I've been watching you all the time," she said. "I thought you must be a humbug from the conceited way you talked. And stealing my clothes in the bargain, too. Well it's the law I'll have on you!"

Toad's temper boiled over, and he lost all control of himself.

"You common, low, *fat* barge woman!" he shouted. "Don't you dare talk to your betters like that! I would have you know that I am a very well-known, respected, and distinguished Toad!"

She made a run at him, and gripped him fast by a hindleg. Suddenly the world turned upside down, the wind whistled in his ears, and Toad found himself flying through the air, revolving rapidly as he went. He rose to the surface spluttering, wild with indignation, thirsting for revenge.

Solid revenge was what he wanted, and running swiftly, gathering his wet skirts in his arms, he overtook the horse, unfastened the tow rope, and jumped lightly on the horse's back, urging it to a gallop by kicking it vigorously in the sides. The barge woman was gesticulating wildly and shouting, "Stop, stop, stop!" but Toad was soon out of sight and hearing.

He had traveled some miles and midday was nearing when his stomach reminded him with a painful growl that he had not eaten. Ahead he saw a dingy gypsy caravan, and beside it a man tending an iron pot, out of which came bubblings and gurglings, and smells—warm, rich, and varied—that twined and twisted and wreathed themselves into one complete, voluptuous, perfect smell that seemed to be the very soul of Nature taking form as solace and comfort.

The gypsy had sized up Toad in a moment as one who would do anything to share the contents of his iron pot.

"Want to sell that horse of yours?" he asked.

Of course, thought Toad, how simple. Get a meal and some money and throw the police off the trail again in the bargain. But before the gypsy would make an offer, no doubt to allow Toad's hunger to further weaken his resistance, the gypsy insisted on reading Toad's fortune. Together they entered the caravan, where a guttering candle glowed dimly from the center of a rickety wooden table. The gypsy drew out a pack of tarot cards and spread them on the table, turning up first the Seven of Swords, which showed a man making off from the field of battle with the swords over his shoulder and a sly look on his face.

The gypsy gazed at Toad intently, and then back at the card again. Finally he spoke.

"I see you in a gypsy caravan. Are you a gypsy?"

"No, no," said Toad. "That was some weeks ago—my passion for the open road before I discovered motorcars."

The gypsy once more gazed intently at him, and then returned to the card.

"An accident," he said.

"Quite right," said Toad, "the caravan overturned."

"Then you steal a car," said the gypsy. Upturning the next card, which showed a bound and blindfolded figure behind a wall of swords. "Eight of Swords. You go to prison," he said.

Toad was about to protest, but the gypsy did not wait for agreement and quickly turned up the next card, which showed a man and woman hurtling down from a lightning-struck tower. "And so you break out of prison," said the gypsy, "and a woman helps you to escape." Fearing that the gypsy might deliver him into the hands of the police, Toad said nothing.

"You steal a horse!" said the gypsy, amazed. "And you steal the same car again! You are a true gypsy," he concluded, "you steal for the pure love of stealing, not for the gain."

Toad blushed quite blue. Now for a toad to blush is quite a feat, because his natural color is green, but a sudden rush of red blood to his face causes a mixing of the red and green, the result being blue.

"What else is in the future?" Toad quavered. "Do I go back to jail?"

The gypsy peered intently into Toad's eyes as if probing the depths of two crystal balls. "No," he concluded. "There will be many honors, many medals. The whole nation will know you for flying an airplane for your country with much bravery."

"What is an airplane?" asked Toad, thoroughly puzzled.

"I am sorry," said the gypsy. "It has not yet been invented."

Finally the needs of Toad's stomach killed his curiosity about the future, and after much haggling it was agreed that Toad would sell the horse for six shillings and six pence and all that he could eat from the pot.

When he had had his fill, he bade the gypsy goodbye, and set off for Toad Hall once more.

It was now mid-afternoon, and in search of Toad, our friends had walked some many miles that day, determined to find him. Recalling the route Toad had taken during his escape, Bertie led them along The River to where it joined the canal, and it was here that they encountered an angry barge woman. Hailing them she inquired if they had seen "a horrid, nasty, crawly Toad. On my nice clean barge, he was. Of course, when I found out what he was, I chucked 'im right into the canal. Served 'im right."

"My dear woman, which *way* did he go?" asked Rat eagerly.

Looking him up and down, the barge woman replied, "That will be the day when I am a Rat's good woman! What bloody cheek! Maybe you could do with a coolin' off in the canal too."

Badger moved ominously between Rat and the barge woman, but before trouble could start, Bertie spoke up.

"Please, m'am, he meant no disrespect. We're trying to find the Toad before he does more harm to poor, honest people like yourself."

"Well, that's another matter then," she replied, temporarily placated. "Say who or what is he? Tellin' me, 'Don't you dare talk to your betters like that,' as if he was some sort of royalty."

Not wishing to reveal Toad's identity, Bertie made up the first thing that came into his mind. "He's escaped from a lunatic asylum."

"I might have knowed," she said. "A Napoleonic complex, that's what 'es got.

Orderin' everyone about, and jumpin' on my horse and ridin' off like that. He thinks 'e's Napoleon, doesn't 'e?"

"A very astute insight," seconded Rat.

"Well, lunatic or not, someone's got to pay for me horse. I can't be pulling a barge up and down this 'ere canal, not with me husband off all the time with the dog hunting rabbits or Lord knows what."

"Dog, did you say?" said Mole looking alarmed.

"He humbugged me, right enough. Dressed up in women's clothes, he was. He ought to be put away! I thought he was a washerwoman. 'Finest business in the whole country,' he says to me. 'All the gentry come to me. Washin', ironin', clear starchin', makin' up gents' fine shirts for evenin' wear.' Twenty girls he said he had workin' for 'im."

Mimicking Toad, " 'Never so happy as when I've got both arms in the wash tub. I simply dote on it.' So I gave him a ride on the barge and he was to do my washin'. So I watched 'im I did. Never washed so much as a dishcloth in his life, I'll lay. And his skin got all crinkly. Then he starts callin' me names, and says he's a very well-known and distinguished Toad. Well, when I had a good look under his bonnet and saw he was a Toad, I chucked him right off, I did. 'Put yourself, through your mangle, washerwoman,' I called out. 'Iron your face and crimp it, and you'll pass for a decent-looking Toad.' That's what I told him. But he jumped on me 'orse and took off. Now then, which one of you represents the Toad asylum that will be compensating me for the loss of me 'orse?"

Just then the gypsy (to whom Toad had sold the washwoman's horse) came around the bend of the road seated atop his dingy caravan, pulled by the very same horse.

"You there, gypsy man! What call have you to be goin' about the countryside with a horse stolen from honest workin' people like meself?!"

The gypsy man appeared startled and reined in the horse. "I paid thirteen shillings for this horse," he said, doubling the amount of the bargain he had struck with Toad. "And if you want him back, it will cost you twenty-six, for he's twice the horse I thought him to be."

"Twenty-six shillings for me own 'orse!" replied the barge woman flabbergasted. "Why he's not worth more than a shillin' a leg. But I'll not pay one shillin', for he was wrongfully taken from me and I'll have him back now!" And with that she grabbed the horse's head and began to remove the bit and bridle. At this the gypsy man jumped down from his perch atop the caravan right onto the barge woman. They went down in a heap and began fighting and wrestling, all the time cursing and crying out, "He's mine, he's mine!"

Mole began sniffling, with Badger refereeing the conflict, while Rat exclaimed, "Toad, you miserable animal. Look what you've done now!"

At that moment, accompanied by his wife, the engine-driver of the train that carried Toad to freedom suddenly appeared. Thinking the barge woman to be the washerwoman he sought, he quickly pulled her from the pile with the gypsy.

"Here now, stop that," he said to the gypsy. "This woman promised to wash my dirty shirts if I'd give her a ride on my train. Against regulations, it was, but no money she had, and 'any amount of kids waitin' at home,' she said, 'playing with matches, and upsetting lamps, and going on generally, the little innocents,' that's what she said. So I says to her, 'what with the shovelling coal and all, it's terribly dirty work, and I use up a power of shirts, and my missus is fair tired of washin' 'em.'"

"Indeed I am," said his missus setting down a heaping hamper of coal-black shirts in front of the barge woman, "And you'll start on them this minute," and with that she grabbed the barge woman by the hair, whereupon the barge woman laid a clout on the head of the missus, and they both went down in a heap, rolling over and over in the dust at the side of the road.

"Oh, dear, oh, dear," wailed Mole.

Then the engine-driver and the gypsy tried to separate the two women and found themselves drawn into the fray. Badger stood at the perimeter, his cudgel at the ready in case the fight showed signs of spreading to his friends.

"Whatever are we to do?" said Rat to no one in particular.

"I suppose it was inevitable that these people would all one day meet," said Bertie, "in one version of the story or another."

"But look what that Toad has got us into," said Badger. "If I had him here now I'd learn him. Let the weasels have Toad Hall, we're better shut of him."

"Now, Badger, I know you don't mean that," said Rat.

"I do, indeed," said Badger with finality.

Less than an hour before, on the high road to town, still wearing his disguise, Toad stepped confidently out into the road to hail an approaching motorcar. In his exalted imagination, riding up to Toad Hall was to be the culmination of his brilliant escape. Suddenly he became very pale, his knees shook and yielded under him, and he collapsed under his bonnet in the very middle of the road. And well he might, the unhappy animal; for the approaching car was none other than the very one which he had stolen before, and in it were the two men and women whose prosecution of his case had landed him in prison.

From under the bonnet of that miserable heap in the road, came these words: "It's all up! Chains and prison again! Dry bread and water. Oh, what a fool I have been! Why did I want to go strutting about the country, singing conceited songs, and hailing cars on the high road, instead of slipping home quietly by back ways! O hapless Toad. O ill-fated animal!"

The terrible motorcar drew slowly nearer, till at last he heard it stop just short of him. Two gentlemen got out and walked round the trembling heap of crumpled misery lying in the road, and one of them said, "Here is a poor old thing—a washerwoman apparently—who has fainted in the road! Perhaps she is overcome by the heat, poor creature; or possibly she has not had any food today. Let us lift her

41

into the car and take her to the nearest village."

They tenderly lifted Toad into the motorcar and propped him up with soft cushions and proceeded on their way.

When Toad heard them talk in so kind and sympathetic a manner, and knew that he was not recognized, his courage began to revive, and he cautiously opened first one eye and then the other.

"Look!" said one of the gentlemen. "She is better already. The fresh air is doing her good. How do you feel now, ma'am?"

"Thank you kindly, sir," said Toad in a feeble voice, "I'm feeling a great deal better!"

"That's right," said the gentleman. "Now keep quite still, and above all don't try to talk."

"I won't," said Toad. "I was only thinking, if I might sit on the front seat there, beside the driver, where I could get the fresh air full in my face, I should be all right again."

"What a very sensible woman!" said the gentleman. "Of course you shall." So they carefully helped Toad into the front seat beside the driver, and on they went once more.

Toad was almost himself again by now. He sat up, looked about him, and tried to beat down the tremors, the yearnings, the old cravings that rose up and beset him and took possession of him entirely.

"It is fate!" he said to himself. "Why strive? Why struggle?" and he turned to the driver by his side.

"Please, sir," he said, "I wish you would kindly let me try and drive the car for a little. I've been watching you carefully, and it looks so easy and interesting, and I should like to be able to tell my friends that once I had driven a motorcar!"

The driver laughed at the proposal, so heartily that the gentleman inquired what the matter was. When he heard, he said, to Toad's delight, "Bravo ma'am! I like your spirit. Let her have a try, and look after her. She won't do any harm."

Toad eagerly scrambled into the seat vacated by the driver, took the steering wheel in his hands, listened with affected humility to the instructions given him, and set the car in motion, but very slowly and carefully at first, for he was determined to be prudent.

The people behind clapped their hands and applauded, and Toad heard them saying, "How well she does it! Fancy a washerwoman driving a car as well as that, the first time!"

Toad went a little faster; then faster still, and faster.

Now as you the reader may or may not recall, in the original *The Wind in the Willows* of Kenneth Grahame, with a half-turn of the wheel the Toad sent the car crashing through the low hedge that ran along the roadside. One mightly bound, a violent shock, and the wheels of the car were churning up the thick mud of a horse pond.

Toad found himself flying through the air with the strong upward rush and delicate curve of a swallow. He liked the motion, and was just beginning to wonder whether it would go on until he developed wings and turned into a Toad-bird, when he landed on his back with a thump, in the rich soft grass of the meadow. Sitting up, he could see the motorcar in the pond, nearly submerged; the gentlemen and the driver, encumbered by their long coats, were floundering helplessly in the water.

Fortunately for the Toad, however, the time when this disastrous event took place was in the *past* in the original story; it lies in the future of *this* story, only moments away, with the Toad on a collision course—one might say—with the event. Now the *place* of the event, the low hedge alongside the road, and the horsepond, lies just beyond the next bend in the road and in this story Toad has just taken the wheel, and begun to accelerate, has heard the warning cry from the gentlemen, "Be careful, washerwoman," and is about to cry out, "Ho! Ho! I am the Toad!" when suddenly as he rounds the bend ahead, he finds the road completely blocked by a quarreling mass of humanity. With nowhere to go he brakes the car only inches from where the barge woman and the engineer's missus are flailing away at one another down in the dust.

Because he sits among such fine folk in the car, he is not recognized by anyone except Bertie. However, as usual, it is the Toad's tongue which gets him into trouble. "Clear the road, you riff-raff!" he shouts, angry at having to rein in his metal steed.

Now as we know, the bargewoman had been abused by the Toad rather recently, and there is something in the tone of that voice which irritates her. Breaking off the combat with the engineer's wife, she peers over the hood to see who has spoken. Recognizing her, Toad quickly lowers his head, the bonnet concealing his face. The two women and the man in the back seat of the motorcar stand up to get a better view of the scene in the road.

Just at that moment, the bargewoman's husband and his dog come out of the wood alongside the road. "Hello, what's this?" he says. His dog eyes Mole, but Badger eyes the dog, and it turns away trying to look uninterested.

"Never here when you're wanted," says the bargewoman to her husband, dusting off her clothes, and indicating the engineer's wife. "This woman has assaulted me."

"Now hold on there," says the engineer.

"I've got rights, too," says the gypsy, "and your missus, if that she be, tried to take my horse."

"It's our horse, right enough," says the bargewoman's husband, "I'd recognize it anywhere."

"Well, I paid a washerwoman thirteen shillings for it."

Hoping to mediate the quarrel, the gentleman in the front seat of the motorcar gets out of the car.

"Do you have a bill of purchase by any chance, old chap?" he inquires of the gypsy.

"Bill of purchase, bill of purchase?" says the gypsy. "What do you take me for?"

"Now look here," says the engineer's missus, "my man here gave this washerwoman a ride on his train because she had no money, and she promised to wash up all his dirty shirts. That was the bargain, and she's got to do it now."

"I ain't no washerwoman," says the bargewoman. "I'm a full-fledged captain of that there barge you see lyin' by in the canal, and me horse was stolen by a nasty toad I chucked into the canal."

At the mention of the Toad, the two gentlemen and their ladies began all talking at once, the gist of their animated conversation being that a Toad of great infamy once had stolen their motorcar and been imprisoned for it.

"Well, they must have put him in the lunatic asylum," says the bargewoman, "because these folks have come to take him back." And she turned to face Bongo and Bertie, Badger, Rat, and Mole.

Slowly Toad raised his eyes to see of whom she spoke, and there were his friends—or so he thought—Badger, Mole, and Rat, and the shamefully punning boy Bertie, whose name-calling—Pigeon Toad!—led to his downfall. O wretched Toad! Had he only controlled himself in the face of that terrible pun he would not have embarked upon a life of crime by stealing from the kind persons who had just given him a lift in the very motorcar in which he now sat. It was all the fault of that Bertie boy

44

and his unauthorized—at least by Kenneth Grahame—appearance in the Toad's story. O what a noble epic of the Toad's sterling character Grahame had been composing until that wretched boy appeared. What valor, mastery, nobility of being! Why the Toad could almost compose the entire book himself, just like his wonderful songs.

While he had been thinking thusly, the two gentlemen and ladies had been conversing on a subject which as soon as he began to pay attention raised his ire considerably.

"Did you say," one gentleman asked the bargewoman, "they have come to take him to the Toad asylum?" The two gentlewomen laughed outloud in a most discourteous manner.

"I wonder if that's the Labor Party's idea," said the other gentlewoman, and the two ladies laughed even harder.

For his part, Toad regarded himself as sane as the next animal, but should he sometime require attention for momentary mania, he did not see why it should be denied him, and his color under the bonnet began to change noticeably.

Casting a cold eye at Rat, one of the ladies in the car remarked, "What can we expect next—Water Rats sitting in the Poetry Chair at Cambridge?" And they all broke into laughter again.

"Now look here!" said Rat, "I've written some very respectable verse!"

"Oh, yes, I believe I've read your Ode to a Swiss Cheese," said one of the gentlemen, and all the other humans began to laugh.

"Let me quote it for you, Cyril," said the other gentleman.

"In homage now
Fall I to my knees,
Before thee, O glorious
And holey Swiss cheese."

Now whereas the two ladies and Cyril, the gentleman who had just spoken, had been standing before in the back of the car, his recitation caused them to grab one another for support, and then eventually collapse with laughter onto the seat. Toad turned round and gave them a look that could kill, but they paid him no mind.

Angrily Badger spoke up. "Now look here, I won't have you abusing my friend Rat."

"Oh, won't you now?" said Cyril. "I suppose we can expect to see you at the Club cadging beers."

The two ladies took up his remark and repeated it several times, drawing out all the syllables, "Badgers cad-ging be-ers."

Mole began to sniffle. "Master Bertie, can't you intervene on our behalf," he said.

"Indeed, I shall, Moley," said Bertie stout-heartedly. "You people are very wrong to be abusing these animals like this. It just shows your ignorance."

"Well, I never!" said the engineer's wife, seconded by the bargewoman.

"You're as rude as the cards in *Alice in Wonderland!*" added Bertie for good measure.

"Little boy!" called Cyril's wife. "Children should be seen and not heard."

"And should speak when spoken to," said the second lady.

Now that they were all attacking Bertie, Mole felt very badly about asking him to take their part, and had to remove his spectacles and wipe his eyes because the tears began to course down his cheeks like rivulets.

"You there, Mole," said Cyril, singling him out. "Have you thought of applying to the Miner's Union?" This remark set them all to laughing again, particularly the train engineer, who countered with his own remark, "You'll be wantin' to join the Welsh Miner's Choral Society."

The Toad had enough. Puffing himself up as large as he could, he began to berate the humans in and around the car, collectively and individually. The bargewoman was closest to the car, and she looked Toad full in the face and said, "And who do we have here?"

The car's occupants had stopped laughing, and one of them replied, "A poor old washerwoman, to whom we gave a lift after she fainted in the road."

"A washerwoman, is it?" said the bargewoman. " 'Finest business in the whole country?'" she inquired, quoting Toad's very words to him.

Toad shook his head.

"Twenty girls workin' for you?"

Toad shook his head again.

"Never so happy as when you've got both arms in the washtub?"

Toad shook his head again.

"No?" she persisted. "Well, I think so," and plucking a large horsefly from the horse's flank, she held it tantalizingly under Toad's nose.

"Of course, if you're not a washerwoman, you might be a Toad."

Toad shook his head again.

"A horrid, nasty, crawly Toad. They say a Toad cannot resist a fly—that the tongue will shoot out and catch the fly in spite of all that the Toad can do."

And now everyone had surrounded the car, and from the rear, Cyril had lain a restricting hand on Toad's shoulder. Toad wriggled in agony, trying to bite his tongue, but in spite of his every effort, suddenly his tongue shot out and impaled the fly in the bargewoman's hand. She made an awful face, wiped her sticky hand on her dress, and raised the other to give a tremendous clout to the Toad, who leapt straight into the air, coming down on the car's hood, where the gypsy, the engineer, and a gentleman all made a lunge at him. But no sooner had his feet touched the car then his legs coiled like springs for a second leap that carried him in the direction of Bertie, who cried out, "Wait Toad!" and then over Bertie's head to the safety of bushes, and from there to the canal's enveloping waters.

Having made good his escape from his pursuers, the Toad crawled out of the canal on its opposite bank. As soon as he had rid himself of the demeaning washerwoman's garb, he became his old boastful self, composing verses to a song which he sang as he walked along.

> The world has held great heroes,
> As history books have showed;
> But never a name to go down in fame
> Compared to the wonderful Toad.
>
> The clever men at Eton and Oxford
> Know all that is to be knowed;
> But none of them know half as much
> As wise and intelligent Toad.
>
> When love in Juliet's breast o'erflowed,
> Who was it had her Romeo'ed?
> Was it a Montague or Capulet?
> No!—the dashing seignor Toad.

And then for his very last adventure, he added one more verse.

The train engineer and his missus,
The fat bargewoman and the gypsy,
The Bertie boys, badgers, moles, and rats.
And even aristocrats in their spats,
All chased him down the road.
Who was it had them snowed?
None other than the all-glorious Toad.

Bongo had spent the afternoon riding around in the backwater pool of The River, sometimes sculling with his paws, on the boat that Badger had whittled for him with his teeth. Quite suddenly the water in front of him began to boil with activity, and Toad surfaced in front of him, festooned with a lily pad and flower, and looking for all the world like the Queen of the May. Bongo, stared and stared, and said nothing. So engrossed in composing odes to his magnificence, Toad had walked right into the River.

Meanwhile, back at the scene of chaos which Toad had left behind him by his sudden departure, the police have finally caught up with all the persons humbugged by Toad on the route of his escape.

"What's all this now?" said the chief constable, pushing his way into the crowd—which had grown bigger— surrounding the animals.

"These here are friends of that Toad that escaped from prison that you're looking for."

"Indeed, we are looking for him," said the chief constable, "and the rest of the country, too."

"Well, you needn't look much further. He jumped into the canal not half an hour ago," said one of the gentlemen.

"Tell me, sergeant," said Cyril, "how could a toad manage to escape from prison, and how could he get this far in one day?"

"That's chief constable—not sergeant," said the policeman.

"Oh, sorry," said Cyril.

"Well, you see, sir, he disguised himself as a woman, a washerwoman we're told, and somehow made his way onto a train. We were closing in on the train, when he must have jumped off and given us the slip."

"That was my train," said the engineer, "but he had me fooled."

"Well, we shall need a statement from you, sir," said the constable.

"And then he humbugged me," said the bargewoman, "and stole my horse. Here, what about that, he gave it to that there gypsy."

"Twenty-six shillings worth, it was to me," said the gypsy.

"Ah, now, we can't have that," said the constable. "Receiving stolen property is a crime, which you gypsies seem to ignore."

"I told you, the washerwoman sold me the horse," said the gypsy angrily.

"Nevertheless, you'll have to give it back to this woman since it rightfully belongs to her, and it would be a good job if you were to move along out of this countryside. Go somewhere else for a while."

The gypsy turned his back and stalked off without the horse.

"Now," said the constable, "can anyone provide me with a good description of this washerwoman/toad?" Immediately, Cyril, his friends, and the engineer and his missus, plus the bargewoman and her husband, all began talking at once.

Bertie, Badger, Rat, and Mole moved away from them out of hearing at the other side of the road.

"Now the Toad's in for it," said Badger, "and serves him right I say."

"Well, if the whole countryside is after him, his friends must stand by him," replied Rat.

"We'd better find him," Mole remarked, "before he tries to go home and those stoats and weasels get him."

"Are you for the secret passageway now, Master Bertie?" inquired Badger.

"No, Badger, it's not necessary, even though I know you would dearly love to engage them in mortal combat once again."

"Once again for the *first* time," said Badger.

"I can tell you, however, Badger, that you made a bang-up job of it," Bertie replied. Indeed, he did, and so did all the others.

What a squealing and squeaking and a screeching filled the air the night Toad Hall was recaptured.

Well might the terrified weasels dive under the tables and spring madly up at the windows!Well might the ferrets rush wildly for the fireplace and get hopelessly jammed in the chimney!Well might tables and chairs be upset, and glass and china be sent crashing to the floor, in the panic of that terrible moment when the four Heroes strode wrathfully into the room! The mighty Badger, his whiskers bristling, his great cudgel whistling through the air; Mole, black and grim, brandishing his stick and shouting his awful war cry, "A Mole! A Mole!" Rat, desperate and determined, his belt bulging with weapons of every age and every variety; Toad, frenzied with excitement and injured pride, swollen to twice his ordinary size, leaping into the air and emitting Toad-whoops that chilled them to the marrow. They were but four in all, but to the panic-stricken weasels the hall seemed full of monstrous animals, grey, black, brown, and green, whooping and flourishing enormous cudgels; and they broke and fled with squeals of terror and dismay, this way and that, through the windows, up the chimney, anywhere to get out of reach of those terrible sticks.

The affair was soon over.

"Well, I daresay we did all right," concluded Badger.

"But it wasn't necessary," said Bertie. "You see the author forgot something, and I remember telling my mother about it, and she agreed with me."

"Come on now," said Rat, "you've hinted at this before, but you've never told us."

"Come on, out with it," reiterated the Badger.

"Well, if you remember the story as I recall it, after he escaped, Toad got on a train driven by that engineer," said Bertie pointing across the road to where the constable was questioning the people about Toad.

"Quite right," said Mole.

"Now that train was soon pursued by another train full of policemen."

"Right again," said Badger.

"So all available police were after Toad."

"Sadly, you are right again," said Rat.

"Then Toad met the bargewoman, stole her horse, and sold it to the gypsy, and was about to steal the same car a second time when he bumped into us on the road."

"But what has all that to do with getting the weasels out of Toad Hall?" said Badger. "And why can't I do as I did before—for the first time, of course?"

"Well, probably you could, and my mother said the author probably conveniently overlooked my point so that he could write that scene. But you've had your fun, Badger, so why not let logic get the weasels out?" said Bertie.

"But what is your point, dear Bertie?" asked Mole.

"Simply this. If someone commits a crime and is sentenced to prison and then escapes, the police try to get him back. And we know a whole train of police were after him, so why should they give up now?"

"Surely, they don't," said Rat.

"Ah, but they do," replied Bertie, "at least the way the author writes the story. After you all help him to retake Toad Hall from the weasels, Toad sends out invitations to a banquet of celebration, and then he sings one last little song. Isn't it logical to expect the police to come for him and take him back to prison?"

"Why, yes, now that you mention it," the Rat agreed.

"Hmm, I never thought of that," said Badger. "I see, so the author kept them away so that we could have our big scene."

"That was my mother's theory," said Bertie. "I just innocently asked her, 'What happened to the police?'"

"But what does it mean for this story now?" asked Mole.

"Just you watch," said Bertie. Crossing the road to the constable, with our friends trooping after him, he walked up to the constable.

"They are the people from the Asylum," said the bargewoman.

"Please, sir," said Bertie, "I believe this woman knows where the Toad has gone."

"How would I know?" protested the woman angrily.

"Think now," the constable said, "when you first met him—before you had words—what did he say about himself; where was he going?"

"Well, now that you mention it, he said, or she said that his married daughter had sent for her, and he wanted to get to Toad Hall; or *near* there, because that's where she lived," replied the bargewoman.

"Constable, may I ask one question of you?"

"Yes, indeed, Mr. Badger."

"I believe I read it in the papers that when Mr. Toad was arrested for theft and cheeking policeman that he was the socially prominent Mr. Toad of Toad Hall."

"Yes, quite right, Mr. Badger."

"Then you are aware that he resides in Toad Hall?"

"Yes, indeed."

"Why is it then that no policeman has yet sought him there?"

"Well, Mr. Badger, I can't rightly say...Hmm, maybe it's because we weren't written that way."

"I dare say," said Badger.

"Well, you must seek him there now," said Bertie, "only be very careful, because he has hired a whole army of stoats and weasels to defend him, and there's sure to be shooting."

"Well, then, we'll call out the army," said the constable.

With the thanks of the police and the crowd ringing in their ears, Bertie and our friends headed up the road as fast as they could go towards Rat's home.

"I must say, I feel as if we had somehow *betrayed* Toad," sniffed Mole.

"Don't you see what Bertie's done," countered Rat, "he's letting the police take back Toad Hall."

"Well, at least we can watch, can't we?" said Badger.

"I wouldn't miss it for all the world," said Bertie. "You took them by surprise. This should be a better fight because the weasels will be ready."

"Why not tell the police about the secret passage, Master Bertie?" asked Mole.

"What and miss such a jolly good show! Where's your fighting spirit, Mole?"

"Then come on lads!" said Badger, twirling his cudgel, "in case we're needed in the second wave."

When they had arrived back at Raty's, they found Toad seated by the fire drying out his skin, which had become all grey and wrinkled from having spent so much time in the water. However, he had found Rat's smoking jacket, and had wrapped this around him. Bongo had brought his boat inside and was playing with it on the floor in front of the fire.

When Rat saw Toad in his best jacket, he became quite angry. "Now washerwoman, out of my house," he said.

"Why Raty, dear chap, don't you know me?" said Toad.

"*I* know you right enough," said Badger, "and should be laying on with my stick."

"A fine way to treat an old friend," sulked Toad.

"An old friend that abandoned us on the high road days ago," said Rat contemptuously.

"Oh, please let me explain," Toad whined. "You don't know what I've been through since I last saw you."

"Oh, but we do know," said Mole. "Young Master Bertie has told it all, some of it before it even happened." Bertie was kneeling on the floor talking to Bongo.

"Drat that wretched boy!" blurted Toad. "If he hadn't said that horrid pun, I never would have embarked on a life of crime."

"That's a lie, Toad, and you know it," said Bertie. "You were going to steal that motorcar whether I happened into your story or not. As a matter of fact, if I hadn't come into your story, you would have stolen it again, because that's what you had in mind when you bumped into us on the road."

"How terribly cheeky of you!" exploded Toad. "How could you possibly presume to know what is in my mind!"

"Oh, cut the gas!" said Badger. "He does know the whole story, even before it happens."

"But I don't understand," said Toad.

"Listen, Toad," said Bertie. "You're going to have to turn over a new leaf. You've taken advantage of your friends too long. And if you don't cooperate, we'll turn you over to the police."

"Turn me in?" said Toad, unbelievingly. "My friends—turn—me—in."

"That's right," said Badger. "And besides, you have no place to go. Toad Hall is occupied by weasels from the Wild Wood."

"Squatters!" raged Toad. "A man goes to jail for a month, and the Labor Party gives his home away!" Toad's color rapidly changed from grey to green to yellow, red, and purple.

"Oh, sit down, Toad, and do try to listen. On our way back here, we worked out a plan that will get you Toad Hall and your freedom," exclaimed the Rat. "And you can thank Bertie. He used his nimble brain to have the police get Toad Hall for you."

And sit and listen the Toad did. The plan called for him to deed Toad Hall to Lord Bertram, Bertie, who would occupy the place with his three friends as custodians. An inquiry would then be made as to Toad's whereabouts, at which the Rat would appear as the star witness. Then, if Toad had stuck to the straight and narrow during this time, residing out of sight in the depths of Mole End, the final stage of the plan would be implemented to permit him to return to Toad Hall. Toad took his oath to abide by the articles of the plan, and all the animals and Bertie swore themselves to secrecy.

In the night, Toad was spirited to Mole End, along with a food supply which included most of Rat's tin salmon, and all of his back issues of "Popular Mechanics," which Toad insisted on taking to while away the time until his grand return to Toad Hall. But our friends had not yet revealed to him the final stages of the plan, and so Rat commented, "It's not going to be as grand as you think, Toad." Then they left him, weeping contritely, and vowing never to see the light of day until they would come for him when the way was made clear.

The next morning they were awakened early by the sound of motor-torpedo launches moving up The River and the clanking of tank treads on the High Road. The army and the navy had been called out. At midday a considerable force was gathered on the heights surrounding Toad Hall. An emissary had been sent to negotiate a surrender, but he could not get within an hundred yards of the Hall before his flag of truce was shot to pieces. Anyone who has had dealings with weasels knows that they are a miserable lot.

By approaching one of the artillery units with watercress sandwiches and hot chocolate, Rat, Badger, Mole, and Bertie were able to learn the commander's plan, and they were invited to view the battle from the safety of the artillery unit's foxhole on a hilltop peak. Canisters of gas were being loaded in the guns of the tanks and artillery.

"Give them mustard and pickle gas," Mole told a corporal, but the army had a plan which provided for ferreting out the weasels with as little damage to the Hall as possible. Truth to tell, the Law intended to put Toad away for life and requisition Toad Hall as an officer's club for the Army and Navy. Therefore, the gas to be used was laughing gas.

Shortly after the order to fire had been given, two tanks rambled up to Toad Hall's venerable windows and fired just two shells each, for upstairs and downstairs, and in no time at all the weasels were out on the lawn collapsing with laughter. Then the artillery dropped their shells neatly around them, creating a thick fog of gas, into which the infantry moved wearing gas masks.

"You've got to admit," said Bertie to Badger, "it beats cudgels and sticks."

For some time after the skirmish between the local constabulary and the stoats and weasels defending Toad Hall, the general populace—both human *and* animal—remained in an uproar, for such pleasant English countryside had never seen the like before of the horrendous events precipitated by Toad's wholly unnatural affection for motorcars, compounded of course by his natural swaggering disposition.

Had he learned his lesson? Only time would tell, but an inquiry was made as to Toad's possible whereabouts by the same local magistrate who had condemned Toad to prison.

Rat proved to be the star witness and his picture and words were in all the papers. According to his testimony, he had one day an extensive conversation with a Norwegian Sea Rat, who was seen by many other of the animals in and about the Wildwood. The gist of the discourse was to convince Rat to give up his comfortable life and snug home in the river bank and to follow wherever the Sea Rat might lead him. And it was only physical restraint by his good friend Mole (who also took the stand to so swear), that prevented him from following the seafarer's track, so eloquent were his talks of Grecian Islands, golden days and balmy nights under strange stars set in a velvet sky.

"But I am afraid Toad was none so lucky, your Honor," said Rat concluding his testimony. "Toad was always, shall I say, a little touched in the head."

"His many manias attest to that," concurred the stentorian voice of Justice.

"I am afraid," said Rat, "the seafarer's words convinced Toad to leave these shores forever."

"And to where was the Sea Rat bound?" asked Justice.

"Constantinople," replied Rat.

"Then we shall never see the infamous face of Toad again. Case closed."

As part of the bargain which Bertie had struck with our friends, they were to aid him as much as possible in finding his purpose, or to provide some clues—at the least—as to why he had been precipitated into their story by the innocent act of his monkey's having opened "their" book.

One night Rat burned the midnight oil very late. After consulting many ancient and worn leather volumes revealing favorable configurations of sun and moon and their accompanying celestial train, he was sure he had found a propitious "Time" in order to aid Bertie. As for the "Place," he consulted one day with Mole.

"Moley, old chap, do you remember the time when Otter's young son Portly disappeared and we went looking for him?"

An uneasy expression came over Mole's face. Removing the tiny spectacles perched near the end of his nose, he rubbed his eyes reflectively. Rat waited for his answer.

55

"I do recall our going out to look in the boat...and I remember being back in the boat...but there is a big gap in between. When I try to remember what happened in between, something *pushes* on my brain as if I'm not supposed to remember. Sometimes I hear music—faraway music that seems somehow to be connected to that place."

"Ah, then you remember *where* we went to look for Portly?"

"Yes...We followed The River until it divided, and then in a backwater somewhere a great dam covered the water from bank to bank. There was a small island in midmost stream, lying as if anchored there, and rimmed by silver birch trees, alder, and willow. I remember Portly being in the boat with us when we left, but then the piping began and it said to forget."

"You're right. We were afraid there, and the music did not want us to be afraid afterwards, but it was lilting, lilting, and *so* beautiful."

"Were there words, Raty?"

"Yes, there were words, but they finished by making us forget...Sometimes I seem to hear the words again, in the wind, the wind in the willows, it's like a harp and the willow limbs are its strings."

"Before I fall asleep at night it drives me nearly mad...*trying* so hard to remember. Oh, Raty, can you remember even *one* word? It would help so much."

"I don't know if I can recall an *actual* word, but I do remember a feeling of a word."

"Oh, do, do tell me," exclaimed Mole.

"That word," said Rat, "is Helper."

Calling in at old Otter's, Rat and Mole found him delighted at their offer to take Portly with them for an afternoon's outing. Since his return on the day he was found by Mole and Rat, he had changed somewhat, sometimes becoming strangely quiet in the moonlight along The River's banks, and cocking his head as if straining to hear a distant music. His father found it painful also when Portly would not respond to questions about where he had been during his disappearance, saying only he could not remember, as if hiding something from his father.

When they had pushed off again and Portly was secure in the bottom of the boat, Rat made the introduction to Bertie. Portly was shy at first, because Bertie was his first human, but before long he had screwed up his courage enough to ask Bertie if he might hold Bongo, and, indeed, became so attached to the little monkey that by the time they had returned to old Otter's dwelling, Portly gave up Bongo only after a great display of tears.

But to the outing on The River. The Willow Wrens were having an angry chat in the dark tapestry of the bank. Some question as to responsibilities of nesting had set them off, and the neighboring jackdaws had chipped in with their two cents worth and more. A gull flown upriver from the sea paused to see if in the exposed nest there might be an egg for the taking. But Jenny and Jim patched the torn seam of domestic bliss, and the gull flew on disgustedly.

When the fork in The River came into sight, Rat asked Bertie to join him in the stern.

"Master Bertie," he said, "do you see where The River divides ahead?" Bertie nodded. "Should we go left or right there?"

Bertie looked puzzledly at Rat, but realized that he was being called on to make an important decision, and then said firmly, "Left."

Further along the way, Rat again challenged Bertie to select the proper backwater. When he had, and they were in sight of the dam, the island swung into view ahead of the bow. Without being prompted, Bertie said to Rat, "We go there!"

As the bow nudged the island's bank, Portly leaped out with a joyous bark. Mole grew silent and kept his eyes on the bottom of the boat, avoiding looking at the island.

Rat placed his paw on Bertie's shoulder, and Bertie clutched Bongo with one arm, forgotten for the moment by Portly.

"It looks dark in there," said Bertie, momentarily afraid.

"Neither Mole, nor Badger, nor I can find your purpose for you," said Rat, confidingly. "Only you can do that. However, the Time may be now, and this—if any—is the Place. Our help is to bring you here. We have been here once before—to find Portly. He will go there *naturally*, without remembering, and you may follow him. Leave Bongo here with me."

Bertie hesitated for a moment, and then got up and stepped quickly onto the island. As if they were about to play some happy game, Portly ran back to him barking invitingly, and then made off towards a dense stand of willows.

Following the bounding otter proved difficult work. But whenever he slipped from sight, his joyous barking could be heard.

Bertie came to a place where a giant willow had grown many limbs parallel to the ground, and from these there hung down a green cascade of shoots and leaves, making a natural bower within.

Suddenly the otter was silent and nowhere to be seen. And just as suddenly, all the birds ceased singing, and the very wind in the willows held its breath. Bertie watched as his hands began twitching; then, for the first time in his life, he began to feel the clammy hands of fear running its fingers through his hair, and drawing a long, cold hand down his back. Something was forcing him to the ground.

Then his mind and intuition began to function once again, and he realized that his fear was not of anything tangible that could harm him, but of the Unknown itself, and that he would have to enter the bower despite the child within him that cried out, Please, oh, no!Don't make me go there!"

At the very center of his being, there was another's voice, a hero's call which told him that within himself he was already a man, and his body would follow after. So he put away the child he was supposed to have been, and parted the green curtain where no path lay and made his own solitary way.

For a moment he regretted—but only for a moment—his intrusion. Before him, over twelve feet high, with his back supported by the willow trunk and his great rump and cloven hooves resting on the grassy sward, his human hands running delicately over shepherd's pipes, pressed to his human mouth, and his head crowned by two long, sickle-shaped horns stood revealed The Great God Pan.

Bertie's intellectual recognition of Him came from a picture he had gazed upon many times in a book of mythology. But the real recognition came in the pit of his stomach, and in his heart which he felt stop its beating. "This is the time and place to die," he felt. "Or, the time and place to be reborn," came the hero's call from within. And then at once in a blinding flash he knew them to be one and the same moment.

Pan's gaze had fixed him from the moment he had parted the green curtain. Now he could no longer face the piercing red eyes that burned into his own eyes like a pair of fiery rubies, and dropped his eyes, as he wanted to drop to his knees in homage.

Around the Great God with eye-lids half-closed or shut as if asleep ran a half-circle of birds, animals, and reptiles of every size, shape, and color, but they were all *English* animals, the animals of *this* Place, and the music was for them, because although the huge chest rose and fell as it filled the pipes with air, Bertie could hear not a sound.

Then he remembered why he had come, the quest for his purpose, and he raised his eyes again to the Awful Presence of Pan. Before he could question Him he had to know One Thing, one very important Thing, and as it formed in his mind ("Are You the God of All-that-is?"), Pan put down His pipes, and a smile played about the edges of His lips, neither confirming nor denying.

But He took his eyes from Bertie and gazed above, as if regarding another realm, and in that moment a Vision formed on the grass at his feet, a Vision that began with a child's pram, and inside it a small boy and a stuffed monkey lying at his side.

Bertie recognized the baby carriage and Bongo at once, but he did not remember himself looking so ugly, like a wizened little old man.

Then behind the pram a garden bench materialized, and Bertie knew that it sat at the very bottom of the garden, the place where his pram had been when the bomb hit.

And then there came slowly into view, sitting on the bench, and rocking the pram with one hand each, while their other hands entwined together, his mother and father. Their faces were clear, and there was a joyous light radiating out from them.

At once he knew in his guts, in his heart, and in his head, his Purpose. Then the Vision began to fade, becoming fuzzy at the edges. All at once, as if something had been forgotten, there was a stirring in the pram, and Bongo sat up and looked out directly into the eyes of the dumbstruck Bertie. With that, everything vanished, including Pan and all the animals around Him.

When Bertie had made his way back to Mole and Rat waiting in the boat with Bongo, dusk's calm had befallen The River, and the shortest night of the year was about to commence.

Ratty was in a kind of daze. He had been hearing the Pan pipes and trying to remember the words that followed after them. They came on the wind in the willows, lingered in the air, and fell about their ears like swallows diving in the darkening sky. The wind played truant with the words. Some stayed for school as Rat told them to Bertie, while others flew far down The River, heard only by the ocean's waves, then flew away forever.

. . . .
Put away the child . . . .   Heard the hero's call . . . .
Parted the green curtain . . . .   And made his own way

It seemed that they had drifted a long way upon The River, hardly speaking to one another, Mole manning the oars from midships, while Rat kept a sharp lookout from the bow, occasionally directing a sharp command at Mole. "Hard a-port!"Or,

"Mind the sandbar to starboard!" Bertie lay curled on the stern plank, Bongo tucked under his chin. The wind had died completely now, and bore no more secrets on the air.

The line of the horizon was clear and hard against the sky, and in one particular quarter it showed black against a silvery climbing phosphorescence that grew and grew. At last, over the rim of the waiting earth the moon lifted with slow majesty till it swung clear of the horizon and rode off, free of moorings; and once more they began to see surfaces—meadows widespread, and quiet gardens; and the river itself from bank to bank, all softly disclosed, all washed clean of mystery and terror, all radiant again as by day, but with a difference that was tremendous. Their old haunts greeted them again in other raiment, as if they had slipped away and put on this pure new apparel and come quickly back, smiling as they shyly waited to see if they would be recognized again under it.

In the distance now, there appeared to be a stately procession of fireflies, moving over a great plain. Mole stopped rowing and turned his head in the direction of the lights, while Rat lifted one leg onto the bowsprit and half-crouched with one paw cocked to his ear. At first only a faint trickle of water could be heard running off Mole's heaved oars, then, as if rushing in to fill the vacuum of silence, there came a kind of humming which rode the stillness up The River until it reached the boat with our friends at the alert.

"That will be the field mice at Stonehenge," said Rat, relaxing his pose in the bow.

"I say, it is the summer solstice isn't it?" said Mole.

Bertie stirred in the back of the boat and sat up, rubbing the sleep from his eyes.

"Let me have a turn at the oars," said Rat to Mole, and, "That's a good chap," as they exchanged places.

Mole took out a clay pipe and lit it, inhaling contentedly the sweet smoke. Then he gazed above to where the stars danced nature's ecstasy.

Now the chanting from Stonehenge could be heard clearer, as The River wound ever nearer to the great plain where the huge monoliths of Stonehenge had been assembled thousands of years before. Bertie stood up on his seat to get a better look. He had seen it before in pictures.

"Rat? Is that Stonehenge?"

Rat nodded.

"Have you been there on Mid-summer Eve's like tonight? They say people were sacrificed there by the Druids. It must be a horrid place."

Mole's eyes twinkled in the moonlight.

"There's as good an example as ever you'd want, Ratty."

"Hear, hear, Moley!"

"I don't understand," said Bertie.

"Why, Man, of course!" said Rat. "Always thinking the worst of every other race and species, while most of the real harm and danger comes from Man himself."

"Flying bombs indeed!" said Mole, seconding Rat's sentiments.

"Now, Master Bertie, Toad has learned his lesson in humility, and you are about to get one from me and Mole here. You were lecturing us a while back about the superiority of knowing what time it is, while we were telling you about natural time. Now one thing the high and mighty human race does *not* know is that Stonehenge is a gigantic timepiece to mark *twenty-one*," and this he said with great emphasis, "positions of the sun and moon whilst rising and setting. Blood sacrifices, indeed!

"And the mice are there tonight, along with many of the animals of these parts, to celebrate their stars. Each is carrying a tiny lantern that is like a star come down to earth. And as they move about in chanting processions passed down for thousands of years, their lantern lights seem to mirror below the movements of their destiny stars above."

"Oh, I like that," said Bertie. "What is a destiny star?"

"The one when you look up that you take for your own."

"Or the one star that takes *you* for its own," added Mole.

"So you see, Master Bertie, you always have a counterpart above, and it may be seen to guide you, and you may take heart that it will survive you."

"If I may play the philosopher again, Bertram, what—I ask you—is a modern watch compared to the intricate calculations and grandeur of Stonehenge?"

Bertie was quite silent in the stern. He looked to Bongo for support, but Bongo just turned away. After all, he was an animal too, like Mole and Rat, and not part of the ignoble race of Man.

Bertie had come down quite a few pegs. He knew now how Toad felt.

"Rat?" questioned Bertie tentatively.

"Yes."

"I saw the most curious thing in the willows on the island. All the animals were gathered, very still, as if half-asleep, but listening to piping which I could see being played—but couldn't hear. The strange thing was a snake lay next to a field mouse, and a hawk next to a wren, and a weasel and rabbit together. How is this possible when normally they eat one another?"

"That music was not for your ears," said Rat. "Humans just see survival of the fittest in nature. They do not see the deeper meaning."

"What is the deeper meaning?"

"It is very difficult to explain, because certain things are known to animals when they are born. It is similar to instinct, but not like a "drive" or compulsion, as humans think of the word.

"The most important idea," continued Rat, "can best be called One-Safeness, but language is inadequate to explain it because this knowing was born within the very first creature on earth. The amoeba knows, the pheasant in the mouth of the fox knows, that its form may be swallowed, indeed, must eventually disappear from the face of the earth, but its essence—perhaps what you humans call the soul—is never in jeopardy, evolves, becomes more, and is never separate from the One, which is All. All of us creatures, including humans, make up the One, which you think of as God. But the One is All, and never judges us.

"So to return to One-Safeness, the One holds intact even the briefest possible life and the most minute form, so no creature can be unsafe or lost, or forgotten. Life seems so complicated, but it is that simple. That is what we animals know," said Rat certainly.

"That must be why I was shown my mother and father—so that I would realize they did not end with the buzz bombs, and so that I could come to know my purpose," said Bertie.

"That must be why you came to us, so that the animals could teach you the lesson of One-Safeness. But now do you know your purpose?"

Bertie put his right hand on his heart, and his left hand on his stomach. "I know it here and here, but in my head I have forgotten it."

"Then you know it certainly," said Rat with finality.

The moon had risen to conjunction with the earth's shadow, and all the destiny lanterns had long since left Stonehenge when Rat turned his little craft hard on one oar and nosed into the landing of his home on The River bank. Mole jumped ashore and tied a line on the dock, while Rat stood gazing at Bongo and Bertie asleep once more in the stern.

"Goodbye, little chaps," he said, and then stepped quickly out of the boat. Untying the loop that Mole had made, he gave a little push and the boat moved again into the quick-flowing stream. "May you get home safely, wherever home may be." And then he turned and disappeared quickly into his house.

When the light went on inside, Mole could still be seen standing on the dock, sniffling into his handkerchief, and gazing after the vanishing craft which was on a river whose course would not be run until it reached a mighty ocean, so strange that in its depths children like Bongo and Bertie, and all manner of wonderful things played, awaiting the moment to be born or reborn again.

And so it was, that after a decent enough time, time enough for things to settle down again after the uproar accompanying Toad's mania, incarceration, and escape, Toad Hall was visited by a "distant cousin" of the infamous Toad, one so unlike the other as day to night. All the animals were impressed by his humility and wondered aloud why Toad's own branch of the family had not been graced by a drop of his humble blood. Indeed, the new resident of Toad Hall went daily about the countryside carrying little baskets of provisions for animals less fortunate than himself, and his calm and cheerful manner made him welcome throughout the countryside. He had one fault, if it could be called that, an unnatural abhorrence of motor cars, for he never could be coaxed to ride in one, and if he were walking on the road and one chanced to pass by, he could be seen shutting his eyes and holding his ears from the sound of the car's horn, poop, poop.

Now at this time in his life, in order to enhance his standing in the community of The Riverbank, this "distant cousin", who was none other than the infamous Toad, sent up to London to have his family tree researched. What came back surprised even Toad, and he immediately sent out invitations to his friends Mole, Rat, and Badger to join him in a great celebration.

No word of the reason for such a grand party had been printed on the gilt-edged cards, so our friends waited patiently throughout dinner for a word from Toad. After all the dishes had been cleared, and a huge bottle of champagne uncorked with a thunderous POP like the backfiring of a motorcar, which momentarily sent Toad's eyes spinning as his motorcar mania stole over him again, Toad was finally ready to spill the beans.

Undoing the red ribbon holding a large, rolled parchment, he sent it unrolling down the middle of the banquet table. It came to rest at the other end of the table next to Badger's glass. "What is it?" he asked gruffly.

"A genealogy!" cried Toad.

"A what?!" said Badger.

"My family tree."

"Dear me," said Rat.

"Going back thousands of years," cried Toad. "What it proves is that I'm not a toad at all, but an amphibian."

"Oh, dear me. What does it all mean?" asked Mole.

"From the Latin *amphibios*, meaning equally at home on land or water," replied Rat smartly.

"You above all, dear Rat," said Toad, "know the joys of The River's aquatic sport, and were the first to introduce me to the joys of punting, which I loved above all else until that abnormal craving for motorcars nearly ruined my life. But don't you see, Rat, it's as clear as the bristles on your nose. I wouldn't have felt so at home in the water unless I were an amphibian."

"I still don't understand," sniffed Mole.

"What Toad is trying so hard to avoid saying is that he is a frog," stated Badger.

"That is the vulgar name, of course," said Toad, "but I prefer amphibian. Don't you see, chaps, everything is quite clear. Toads don't like water, preferring to commonly lay about in dust, or to wallow in mud at best, but frogs love the water."

And here Toad broke into rapturous songs of his own composition.

O for the life of the lily pad
There is none better to be had.
Even in the rain, it's not so bad,
You don't need clothes—your back's moss-clad.
O for the life of the lily pad.

His friends had not seen Toad so gay since before the days of his imprisonment, so they said nothing to discourage him. Soon thereafter Toad set workmen to erecting signs about the premises of his home, so that Toad Hall was transformed into Amphibian Farm, and Toad changed too, his eyes misting over at the prospect of raising pedigreed sea horses.

Of course, Toad's new pre-occupation lasted no longer than those that had gone before: punting, the gypsy caravan, and motorcars, and was replaced in time by a new mania, so complete in its hold upon him that the others paled in comparison and were forgotten: flying!

One lazy summer afternoon there appeared in the sky above Toad Hall—excuse me, Amphibian Farm—an old-fashioned two-winged aeroplane, and Toad responded as the moth to the flame. As its pilot put it through its paces, performing a series of loops and barrel-rolls, diving and zooming, Toad's eyes once again as in days of yore grew wide as saucers and the pupils began to spin as rapidly as pinwheels, until he fell down frothing at the mouth.

When he came to his senses, nothing would do but that he have his own plane, and soon he became the terror of the countryside, buzzing the cows at pasture and curdling the milk in their udders, zooming over the henhouses so that they laid no more and the cocks, too timid to crow, gave up henning. The countryside was in an uproar.

Rat, above all, hated Toad's new mania. At the first crack of dawn, he would awaken to a distant drone. Then louder and louder it became. Toad's bi-plane skimming over The River's water, would loom out of the darkness and pull up just over Rat's roof, rattling the windows and propelling Rat out of bed in a rage.

Toad swung his plane into a slow banking turn, throttled down, and glided low over the agitated Rat, who stood shaking his fist at Toad.

"Dawn patrol!" called out Toad, leaning out of the cockpit.

This chapter of the many manias of the nefarious Toad has a happy ending, however, for the time now is the last few weeks of peace before unpleasantries were to commence in World War I, and soon Toad's aerial talents were to be put to good use in the service of his country. He was in fact the only Toad—excuse me, amphibian—in the war to win the Distinguished Flying Cross, and Germany's Ace of aces, the notorious Red Baron, fell in a dogfight at the hand of the Flying Amphibian, as Toad came to be called in the newspapers that documented his exploits. "FLYING AMPHIBIAN FLAMES BARON," ran the headline. After the war, he was to be knighted by the Queen, and so at long last became a credit to his friends Rat, Mole, and Badger.

As for Bongo and Bertie, they became a memory so faint none of the animals could remember, and disturbed our story no more than the gentle wash of a wave along The River bank.

65

GLOSSARY
BOOK ONE

abhorrence: intense dislike
atavistic: reverting to the characteristics of an ancient ancestor
beating about the bush: not getting to the point
big bang: the explosion of matter wherein the universe began expanding ten to twenty thousand million years ago
bijou: small and well made
black hole: a collapsed star with all of its matter compressed into zero volume, resulting in gravity so strong that not even light can escape, hence a very dark place
bower: a leafy enclosure
cacophonous: having a harsh sound
cacophony: a harsh discordance of sound
cadging: begging
caricature: an exaggerated representation of a person or thing
cerulean: blue
cheek: insult
claustrophobic: so close or narrow as to inspire fear.
commandeered: took over
configurations: positions of stars or planets in the sky
constabulary: police
cudgel: club-like sticks
debonair: carefree or charming
decorum: proper or dignified behavior
demeaning: lowering or debasing, indignity
disporting: amusing one's self
embellishment: an ornamental display or decoration
emissary: messenger
encumbered: burdened
enhance: to intensify
ergo: therefore
eureka!: from the Greek word herueka meaning 'I have found it' an exclamation of triumph at discovering something
felicitous: happily well suited
festooned: decorated with garlands
gesticulating: gesturing or using the hands to express one's self
gunwale: the upper edge of the side of the boat
halberds: long poles with spikes or blades, weapons
horrendous: dreadful or horrible
implemented: performed or carried out
incarceration: imprisonment

invective: abusive words or accusations
mangle: a machine for squeezing by rollers
mania: craze
metaphysical: abstract thought beyond physical causes
monograph: a piece of writing
monoliths: huge upright stones
Napoleonic: behaving with exaggerated self importance
nefarious: wicked
nurturing: encouraging
panorama: a wide view
pedigreed: pure bred
phosphorescence: a glow
physicist: a scientist who deals with physical causes of matter or energy
piebald: colored differently
placated: calmed or quieted
pounds: English money
precocious: smart for one's age
propitious: favorable
provocative: interesting
punting: moving a small boat by pushing a pole against the river bottom
raiment: clothing
requisition: demand
reviling: to verbally abuse or berate
rustic: country-like or rough
sacked: fired or dismissed from a job
shillings: English money
skittles: old-fashioned bowling
solace: comfort
stentorian: powerful or loud
stipulation: condition
subterranean ring: underground sound
tendrils: vines
truncheons: clubs carried by policemen
uncommunicative: reserved and unsharing
verdant: leafy green
voluptuous: pleasurable
wainscot: the wood paneling of the lower part of a wall
warders: guards
wizened: shriveled or dried up

BOOK TWO

Water

Do as you would be done by.
Be done by as you did.

Bongo and Bertie were now quite amphibious. You do not know what that means? You should have learned that word in Toad's story when he found out that he was not a toad but a frog. The truth is Bongo and Bertie now had growing along the back of their shoulders the loveliest little collars of lace that you have ever seen. But they were not lace frills. Oh, no, they were gills.

Now the wonderful things about gills is that they collect oxygen in water so that a creature can breathe. Fish have them naturally, but how Bongo and Bertie came by them I am sure I do not know. Nor do I know how they came to find themselves swimming in the sea, when only a few pages ago they were moving down The River towards the sea in a wonderful little boat made a gift to them by Rat.

One thing, however, was certain. They were not in the same story anymore. Can not you tell? The style has changed. This story has a bit of a Victorian ring to it. But do not fret; it becomes much more lively as it moves along.

I am the new author, and am a clergyman, and I shall attempt to convince you to believe the one true doctrine of this wonderful fairy-tale; which is, that your soul makes your body, just as a snail makes his shell...instead of fancying, as some people do, that your body makes your soul as if a steam-engine could make its own coal; or, with some other people, that your soul has nothing to do with your body, but is only stuck into it like a pin into a pincushion, to fall out with the first shake. For the rest, it is enough for us to be sure that whether or not we lived before, we shall live again.

But enough of moralizing. Bongo and Bertie were having a wonderful time. Being the little monkey he was, Bongo of course found new ways of playing in the water. Catching a wave just before it broke, he would ride down down the crest, controlling the direction he took by dipping a shoulder left or right, while his tail acted like a rudder anchored in the back of the wave.

What he was doing is now called body-surfing, and some of you may think that surfing was invented in the twentieth century, but you would be utterly wrong, since this story was written in the nineteenth century. Therefore, the logic is inescapable: Bongo invented body-surfing.

Soon he taught Bertie the hang of it, and within an hour's time they had refined their sport to swooping back and forth behind one another as they came whooping down the waves, Bertie sometimes riding right over Bongo's tail, until they often collided in an explosion of laughter and spray.

You see, a Victorian book is not all moralizing and can have a bit of fun in it. One time an otter poked his head out of the water right between them. Taking him for their old friend the otter of the last story, Bertie inquired of their whereabouts.

"I'll leave you to find that out," replied the otter, ducking beneath the waves and darting away.

Bongo and Bertie continued their sport, and presently the otter returned.

"Have you found out where you are?" he asked, surfacing between them.

"How could we," said Bertie. "Nothing has happened since you left."

"Well," said the wily otter, rolling his eyes, "I did say I'd leave you to find out."

He disappeared, and in a few moments surfaced again, noticeably bigger than before. "I left. Have you found out?"

"Holy, Moley," said Bertie, "I hope this isn't *Alice in Wonderland*. Lewis Carroll was a mathematician, but not a physicist like Bertrand Russell, whom *I'm* named after."

"We do not have a jabberwocky," replied the otter, "but we do have the last Gairfowl."

"The last *what*?!" said Bertie.

"Gairfowl. If you go to Shiny Wall on the way to the Other-End-of-Nowhere, you will come to the last of the Gairfowl, standing on the Allalonestone, all alone."

"A little illogic goes a long way," replied Bertie, "and I can't stand puns."

"Of course you can't stand puns," retorted the otter. "They fall down every time. But don't kid me, kid, or try to get my goat, because it was that little pun of yours that sent Toad to jail."

"How do you figure *that*?" demanded Bertie.

"Welllll," said the otter, "Toad got *so* mad at your calling him *Pigeon* Toad that he ran amuck. One little pun was all it took. Suppose you had resisted saying it and helped out instead by *warning* the others about his stealing a car. Then no law to deal with. No weasels in Toad Hall, and Toad reformed."

"But that's just it," countered Bertie. "Toad *wouldn't* have reformed, if he *hadn't* undergone his trials and tribulations. And besides, he ran amuck in the first story anyhow; so I don't see how I can be blamed for his bad fate. Now look here, you're not the same nice otter as in the last story, so how do you know about me? And stop growing bigger like that!"

"The fairies told me about you. They know everything. And they're expecting you."

"Expecting me *where*?" said Bertie.

By now he and Bongo were no bigger than the eye of the otter, and the otter had to stare at him in order not to lose sight of him.

"Everywhere," replied the otter, "and anywhere. That's where the fairies are."

Some people think there are no fairies. Well, perhaps there are none in Boston, U.S. And Aunt Agitate, in her Arguments on political economy, says there are none. Well, perhaps there are none—in her political economy. But it is a wide world, my little man—and thank heaven for it...and plenty of room in it for fairies, without people seeing them; unless, of course, they look in the right place. The most wonderful things in the world, you know, are just the things which no one can see. There must be fairies; for this is a fairy-tale: and how can one have a fairy-tale if there are not fairies?

You don't see the logic of that? Perhaps not. Then please not to see the logic of a great many arguments exactly like it which you will hear before your beard is grey.

But what of Bertie and Bongo and the rapidly growing otter? Well, Bertie had failed to figure out what was happening to him. How could Bertie become a scientist when failing to understand one of the first lessons of spatial relations. When one object gets *bigger* than another, it does not necessarily mean that it is *growing*. No, indeed. The other object may be shrinking, and that is precisely what had happened to Bertie and Bongo.

Ah, now comes the most wonderful part of this wonderful story, because now they were precisely 3.87902 inches long, just the size the fairies make someone when they turn them into a water-baby.

A water-baby? You never heard of a water-baby? Perhaps not. That is the very reason why this story was written. And no one has a right to say that no water-babies exist, till they have seen no water-babies existing; which is quite a different thing, mind, from not seeing water-babies; and a thing which nobody ever did, or perhaps ever will do. You must not talk about ain't or can't when you speak of this wonderful world around you, of which the wisest man knows only the very smallest corner, and is, as the great Isaac Newton said, only a child picking up pebbles on the shore of a boundless ocean.

The fairies, who had been using the otter to entertain us with his light patter, while they readied Tom in the wings, now grabbed the hook. Dutifully he disappeared under the waves without so much as a "thank you, mam."Who is Tom, you ask? Why Tom is the little chimney-sweep about whom I wrote *my* story. And as if on cue he swam up to Bongo and Bertie. For several days now Tom had been swimming about having many adventures on his own, but this was the first time he had seen anyone who looked the least bit like himself.

"Hello," said Tom to Bongo, "you must be a water-monkey." Bongo nodded enthusiastically.

"And you," said Tom to Bertie, "must be a water-baby."

"Certainly not!" responded Bertie somewhat angrily. "As anyone can plainly see, I am a boy!"

The truth is they became the very best of friends from that first meeting because each had qualities which the other lacked, and so they were a well-matched pair. Opposites attract, or so the philosophers say. Bertie dubbed them The Three Musketeers, and they immediately took an oath to stick together for life. As they swam along, Bertie told Tom his story so far, and then it came Tom's turn.

Before he became a water-baby, Tom lived in a great town in the country north of London, where there were many, many chimneys to sweep, and plenty of money for Tom to earn and his master to spend. He was apprenticed to his master—a blasphemous drunkard named Grimes—at the age of five, since orphans then were put to work to earn their keep.

Now you may think that Tom became a water-baby because he never bathed, which is true enough, but you would be quite wrong. Tom could not read or write, nor cared to do either, but he did not not wash out of ungodliness; for cleanliness is indeed next to godliness, but because there was no water in the court where he lived. He never had heard of God, except in words which Grimes used when his mood was angrier than most, and he had never been to Sunday school, or even a Monday through Friday school.

He cried half the time, and laughed the other half, which he regarded as a fair shake from the world. He cried when he had to climb up inside the dark chimneys, scraping raw his knees and elbows. And when he rubbed the inside of the flues, reaching up with his wire brushes to the narrowest part near the top where he could not climb, the soot descended on him, blinding his eyes. That is when he cried hardest of all, for he soon discovered that his tears washed his eyes clean, thus Tom was able to find some good in most things.

"Bongo could have done that for you," said Bertie, who had never seen boys his own age forced to clean chimneys in London. Bongo shook his head. He was not interested in cleaning chimneys. He was particularly fastidious about his cream-colored fur, and when Toad had placed his muddy paws upon the prow of the boat Badger had whittled for him, it was all Bongo could do to resist pushing Toad off the boat and under the water.

"Monkeys could be trained for the job," persisted Bertie. "They're smaller and better climbers." Bongo shook his head *very* emphatically this time, but Bertie was not looking.

Tom continued his story. When he was not crying, he was laughing, pitching pennies with the other boys, leap-frogging over the posts—toads could be trained to do it better, thought Bongo—or rolling stones at the horses' legs as they trotted by, provided there was a wall nearby to hide behind. As for chimney-sweeping, and going hungry some days, and being beaten by Grimes in the bargain, Tom took it all as the way of the world, like snow and hail, thunder and lightning, and stood manfully with his back to it till it was over, as his old donkey did to a hailstorm; and then shook his ears and started laughing again; and thinking of the good times ahead when he would be a man and sit like Grimes in a public-house and drink beer from his own mug, and smoke a beautifully carved pipe, and nod his head wisely from time to time at his old playmates, who would now be grown like himself; and he would have not one but two apprentices, so that he never would have to climb a chimney again, and he would be kind to them and never bully just because he was bigger, and at the end of each day he would ride forth through the town on his donkey with his two apprentices behind, like a king at the head of his army, and all the people would watch and say, "There goes Tom, the Master-sweep."

"Perhaps Bongo and I could work for you," said Bertie, interrupting Tom's story.

"No," said Tom, "by then you'll be grown up too."

But the best part of Tom's story he saved for the last. Because at the end of each day, and especially on Sundays when he had finished his pipe and drained his mug to the dregs, he would go home to his own little house where there waited for him a beautiful woman, none other than his own sweet wife.

Tom smiled and rolled over on his back, floating in the sun.

Bertie thought out loud, "A beautiful woman just like my mother." Then he was sorry he had said anything. Tom rolled onto his side and gave him a funny look, because when they first had met, Tom had told Bertie he was an orphan, and Bertie had replied, "So am I."

Bertie decided to change the subject, so he asked "How did you become a water-baby?"

Tom frowned for a moment. In his story he had reached the future when he was a man and free of Grimes forever, and now he would have to go back in his mind again.

"One day," said Tom, "a man told me that Grimes the master-sweep was to go to Sir John Harthover's Place—as it was called—for to sweep his chimneys. Now Harthover Place was a grand place, indeed, even for England's rich North country, with miles of game preserves full of deer, fox, pheasant, and salmon, which Grimes poached at times. Sir John Harthover was a grand old man, whom even Grimes respected, for he could send Grimes to prison once or twice a week for poaching salmon. Not only did he own all the land about for miles, but he was the jolliest, most honest, sensible squire as ever kept a pack of hounds, who would never fail to do right by all his neighbors. Ah, but truth to tell, Sir John's greatest pride came not from his manorial home or his extensive lands, but from his own darling daughter—Ellie, a mere lass of seven.

And as it was that at three o'clock on that very next morning, Grimes and Tom got up from bed to set out for Harthover Place and to sweep clean its many chimneys. Now, I dare say, you never got up at three o'clock on a mid-summer morning, but I assure you that it is the pleasantest time of all the twenty-four hours, and all the three-hundred and sixty-five days.

Grimes rode the donkey in front, and Tom with the brushes walked behind; out of the court where they lodged, and up the cobble-stoned street where the donkey's hooves rang like hammers on the anvil of early morning. All else was silent. For old Mother Earth was still fast asleep; and, like many pretty people, she looked still prettier asleep than when awake. The great elm trees in the green-gold meadows were fast asleep above, and the cows fast asleep beneath them. Nay, the few clouds which were about were fast asleep likewise, and so tired that they had lain down on the earth to rest among the tops of the alder trees by the stream, waiting for the sun to bid them rise in the clear blue overhead, and go about their day's business of bringing shade or showers to the fields below.

They passed through the coal-miners village, all shut up and silent now; and through the turnpike; and then they were out in the country, plodding along a black dusty road between black slag walls, with no sound but the groaning and thumping of the coal-field's engines. But soon the road grew white, and the walls likewise; and at the wall's foot grew long grasses and gay flowers, all drenched with dew, and they heard, high up in the air, the skylark saying his prayers, and the warbler in the sedges, where he had warbled all night long.

On they went; and Tom looked, and looked, for he had never been so far out into the country before; and longed, and longed to get over a gate and pick buttercups, and look for bird's nests in the hedges, but Grimes was all business, and would have none of it.

Soon they caught up with a poor Irishwoman, trudging along with a bundle at her back. She had neither shoes nor stockings, and limped along as if she were tired and footsore; but she was a very tall, handsome woman, with bright grey eyes, and she took Mr. Grimes' fancy so much, that when he rode up beside her, he called out:

"This is a hard road for a comely foot like that. Will ye up, lass, and ride behind me?"

"Ah, sure, thank you kindly, but I'll walk with your little lad here."

So she walked beside Tom, and talked to him, and asked him where he lived, and what he knew, till Tom thought this must be what a mother does with a son to make him feel at home in the world. Then she asked him, at last, whether he said his prayers, and seemed sad when he told her that he knew no prayers to say.

Then he asked her where she lived; and she said far away by the sea. And Tom longed to know about the sea, which he had never seen. And she told him how it rolled and roared over the rocks on winter nights, and lay still in the bright summer sunlight, for the children to bathe and play in it; and many a story more she told him, till Tom longed to go and see the sea, and bathe in it likewise.

At last, at the bottom of a hill, they came to a spring:not such a spring which soaks up out of white gravel in the bog, among pink bottleheath and sweet white orchids, nor one which bubbles up under the warm sand bank by the great tuft of lady ferns, and makes the sand dance reels at the bottom all the year round: not such a spring as either of those, but a real North country limestone fountain, like those in ancient Greece, where the nymphs and muses sat cooling themselves the hot summer day. Out of a low cave of rock, the great fountain rose, bubbling and gurgling so clear that you could not tell where the water ended and the air began, and ran away under the road, a stream large enough to turn a mill-wheel, among blue geranium, and golden globe-flower, and wild raspberry, and the bird-cherry with its tassels of snow.

Here Grimes stopped, and knelt down, and with his ugly head, fouled the fountain by dipping his head in again and again—and very dirty he made it.

When he had finished, poor little Tom said, "I wish I might go and dip my head in."

"Thou come along," said Grimes, "what dost thou want with washing thyself? 'Twasn't for cleanliness I did it, but for coolness. Thou did not drink a gallon of beer last night, like me."

And here, naughty Tom said, "I don't care what you say," and ran to the fountain and began washing his face.

Tom's defiance was not long in provoking a response from his master, who rushed at him and began to beat him until the Irishwoman cried out. "If you strike that boy again, I shall tell the authorities what I know of you and what happened in Aldermire Copse, by night two years ago."

Grimes seemed quite cowed, and got on his donkey without another word.

"Stop!" said the Irishwoman. "I have one more word for you both; for you both will see me again, before all is over. Those that wish to be clean, clean they will be; and those that wish to be foul, foul they will be. Remember."

And one moment she was there, and the next she was gone, as if into thin air.

Now Tom and Grimes thought long on this the rest of the way to Harthover Place, but neither spoke of it to the other. They were received there by the ash-boy, yawning horribly, and then led down a passageway where the housekeeper met them, in such a flower chintz dressing-gown that Tom mistook her for My Lady Harthover herself. And then the housekeeper turned them into a grand room, all covered up in sheets of brown paper in anticipation of their coming, and bade them begin. And so, after a whimper or two, and a kick from his master, into the grate Tom went and up the chimney.

How many chimneys he swept I cannot say: but he swept so many he got quite tired, and puzzled too, for they were not like the town flues to which he was accustomed, but such as you would find in old country houses, large and crooked chimneys, which had been altered again and again, till they ran into one another. So Tom fairly lost his way in them; not that he cared much for that, though he was in pitchy darkness, for he was as much at home in a chimney as a mole is underground, although not as much at home as Mole of our last story, who found Mole End very comfortable, indeed. At last, Tom coming down as he thought the right chimney, came down the wrong one, and found himself standing on the hearth-rug in a room all dressed in white, with just a few lines of pink here and there.

And then, looking toward the bed, Tom saw under the snow-white coverlet, upon the snow-white pillow, the most beautiful girl he had ever seen. This was none other than Ellie, Sir John's own pride and joy. Her cheeks were almost as white as the pillow, and her hair was like threads of gold, and Tom stood staring at her as if she had been an angel out of heaven.

And so he stood for the longest time, until he saw in the corner across from him the ugliest black little ape that he had ever seen. Now you may well imagine what a shock this was, for only moments before he had beheld the most beautiful sight he had ever seen, and now to have it followed so closely by the *ugliest* thing he had seen was indeed a shock. But an even greater shock was to follow. What did such a little black ape want in that sweet lady's room, desecrating the alabaster purity that was her sleeping quarters?Would she not be frightened if she awoke in the presence of such a sooty hyena?

And so in that moment, Tom determined to be her champion by ridding the room of the obnoxious presence. Making for it he found that it mimicked his movements, and Tom came face to face with the reality of his blackness, for it was himself reflected in a great mirror, the like of which he had never seen before.

And Tom, for the first time in his life, found out that he was dirty; and burst into tears with shame and anger; and turned to sneak up the chimney again to hide his loathsome self; and instead toppled the fire-irons into the fender with a noise like ten thousand tin cans tied to ten thousand mad dogs' tails.

Up jumped the little white lady, and seeing Tom, screamed as shrill as any peacock. In rushed her stout old nurse from the next room, and making up her mind at once that he had come to rob, plunder, destroy, and burn, dashed at Tom. Out the window jumped Tom; scrambled down a giant tree; and ran as fast as his little legs could carry him—away!

Now Tom ran on the rest of that day and most of the next. And when he stopped to rest, he fell fast asleep; but he did not sleep long, for he tossed and kicked about in the strangest way. He dreamed that he heard the little white lady crying to him, "Oh, you're so dirty; go and be washed." Then he heard the Irishwoman saying, "Those that wish to be clean, clean they will be." And then he heard church bells ringing so loud, close to him too, that he was sure it must be Sunday. Whereupon his dreaming self rose up from his sleeping body to find the church and see what it was like inside. But the people blocked the door, and would not let him come in, all over soot and dirt as he was. "Go to the river," they all cried, "and wash before you can come in." And he said out loud again and again, though being half asleep he did not know it, "I must be clean, I must be clean."

And all of a sudden he found himself in the middle of a meadow, with a stream before him, saying continually, "I must be clean, I must be clean." He had got there on his own legs, between sleep and wakefulness. But he was not a bit surprised, and lay down on the bank and looked into the clear, clear limestone water, with every pebble at the bottom bright and clean, while the little silver trout dashed about in fright at the sight of his black face, just as the little white lady had done; and he dipped his hand in and found it cool, cool, cool; and he said, "I will be a fish; I will swim in the water; I must be clean, I must be clean."

So he pulled off all his clothes, and put his poor, hot, sore feet into the water, and then his legs, and the further he went in the more the church bells rang in his head. And then his head went under, for he did not know how to swim, and all the bells were silent.

He never saw the Irishwoman, but just before he came to the riverside, she had stepped down into the cool, clear water at that very same spot; and her shawl and her petticoat floated off her, and the green water-weeds floated round her sides, and the white water-lilies garlanded her head like a halo, and the fairies of the stream came up from the bottom and bore her away downstream in their arms; for she was the Queen of them all, and of much more besides.

"Where have you been?" they asked.

"I have been soothing sick folks' fevered brows, and whispering dreams of well-being into their ears; opening cottage windows to let in the healing fresh air; coaxing little children away from playing in gutters where fever breeds; turning women away from the gin-shop door, and staying men's hands as they were about to harm their wives; doing all I can to help those who cannot help themselves, and little enough it is, and weary work for me. But the good news is that I have brought you a new little brother, and watched him all the way here."

Then all the fairies laughed for joy at the thought that they had a little brother coming.

"But mind, one and all, for he must not see you yet, or know that you are here," said their Queen. "For he is but a savage now, and like the beasts which perish; and from the beasts which perish he must learn. So you must not play with him, or speak to him, until he has some lessons learned:but only keep him from harm's way."

And so now you know Tom's story, and why he has seen no water-babies, for their Queen has forbidden it; yet he has learned of them from the beasts of the sea. Now it is no accident that Bertie and Bongo have crossed paths with him, unless you believe with the new fool-o-sophers of the day that nothing in this world or the next has any purpose. But I am not writing such a story; no, indeed, for to write about nothing is to waste the very ink and paper with which it is written.

Instead you may surmise that Bertie's purpose and Tom's somehow intertwine, and you would be right, clever little reader that you are, although so far, Tom knew only that he wanted to be clean, and once clean, contented himself with having a good time in the sea, for which we cannot blame him too harshly, since he had been climbing up to sweep chimneys almost from the moment he could walk.

"So the fairies made you a water-baby instead of letting you drown," said Bertie when Tom had concluded his story. "Then that must be what happened to us, Bongo. There must be many, many water-babies in the ocean, as many as stars in the sky."

"Yes, there must be," agreed Tom, "but I haven't been able to find a single one. Only yesterday a poor sick fish, who said she came all the way from the Carolinas, told me she was helped by them not once but twice."

"They must be hiding," decided Bertie,"else one of us would have met them."

"But why hide from us if we're water-babies too?" argued Tom.

"Maybe they are different from us in some way."

"Maybe they are afraid of Bongo. He *is* different because of his tail," said Tom.

Bongo shook his head. He knew the water-babies were not afraid of him.

Well, two more unlike companions than Bertie and Tom you never could find. I have told you that opposites attract, so hear now how they differed. Tom, as we know, always wanted to be clean, but truth to tell he was black as Bertie's shadow, covered with chimney soot from head to foot, until that moment when he slipped into the stream and became a water-baby.

When Bertie lived with his mother, before he went to his aunt's, he had a bath *both* morning and night. He did not want to be all *that* clean, but his mother insisted upon it. So you have one who wanted to be cleaner, and could not, and the other who wanted to be less clean, and could not. Despite all the weeks in the water there was still a collar of ingrained grime around Tom's neck that gave away his earlier calling of chimney-sweep.

Above the neck they were quite different also. Although he was a bright lad, and had he been given the chance might have made something more of himself than sweep, Tom could neither read nor write. Bertie, however, certainly was a budding genius. He made connections between things that other children just didn't see. If one was ever destined to find the missing link in Einstein's Unified Field Theory it was probably Bertie. Whereas Bertie was always thinking about the meaning of things, Tom lived instinctively like a little animal, with never a thought in his head except to have fun.

Each balanced what was lacking in the other, and so they became a well-matched pair, with nothing in common except being orphans.

Tom taught Bertie how to have fun, which Bongo had been trying to teach him all along, and Bertie awakened Tom's mind to the fun in finding the meaning behind things. Sad to say, however, their mutual progress did not develop in the most noble direction. Bertie went too far in Tom's direction and became like a little animal, acquiring—sadder still—some of Tom's worst traits, as I shall describe for you herewith.

Pebbles in anemones' mouths: Now if you have never seen an anemone before, you should look carefully the next time you go to the beach. Have mother or father, or both (if you are luckier than the two orphaned boys of our story), hold your hand so that a wave doesn't knock you over. Down under the water, stuck fast to the rocks, live the anemones. When you first see them, you will think you are looking at beautiful blue-green flowers, but their many petals are really tentacles, like the long arms of the octopus. The anemones are the flowers of the sea, even though they are really animals.

Why, wise men of old said that everything on earth had its double in the water. There are land-babies—then why not water-babies? Are there not water-rats, water-flies, water-crickets, water-crabs, water-tortoises, water-scorpions, water-tigers and sea-bears, sea-horses, and sea-elephants, sea-mice, and sea-urchins, sea-razors and sea-pens, sea-combs, and sea-fans; and of plants, are there not water-grass, and water-crowfoot, water-milfoil, and so on, without end?

But the anemones are not plants but animals, and they leave their "petals" as far-flung as possible, and anything that touches them causes the tentacles to draw that thing in towards the center where the mouth is. Now since the anemone cannot move, it has to take whatever comes along. And this is where naughty Tom led Bertie astray, because before meeting Bertie, Tom whiled away many a lonely hour dropping pebbles onto the anemones to watch them close up. No, I do not know whether they ever managed to actually swallow the stones, but I do know if they did they gathered no nourishment. Who knows but that these senseless pranks killed these lovely sea-creatures?But Bertie, who had been a proper little man had quite lost his senses and become under Tom's teaching a boy again, and a naughty one at that.

I should not tell tales on them, but that is what you must expect when the author is omniscient, which means knowing all things, just as God knows all things.

Another nasty game they played was designed to frighten the little fish. Naturally they did not try it on bigger fish, because they were afraid of being eaten. What they did not know is that the fairies protect all their water-babies, so they had nothing to fear.

Here's how the game went. Hiding behind a sea-fan, or a kelp's undulating branches, they would wait until an unsuspecting sprat swam up and then leap out, crying "Boo!" The poor fish spent the rest of the day with its heart in its mouth, even as you and I would if someone had played the same trick on us from a darkened hallway.

But perhaps their worst trick involved the tiny fiddler crabs that scuttled over the many sea-pools in the rocks. Tom and Bertie would collect pebbles from the bottom, and then swim up to sit on a rock to watch for the crabs. When one came along, they would rain the stones off their shells until they abandoned the pool for the sand below. Burrowing down with only their stalk eyes exposed, the crabs would think themselves safe at last, only to find a hail of pebbles flying down from above. This antic inspired the greatest mirth in the two boys.

But what of the third musketeer? Bongo would have none of it. Perhaps because he was an animal himself, even though a *stuffed* one, he did not join in tormenting other animals. So when Tom and Bertie shied the pebbles down, he turned his back on their "fun" and sat watching old Mr. Sun as he made his ancient way across the horizon.

And an odd thing happened at such times. Bongo's face began to redden. Now it is a well-known fact that a stuffed animal cannot obtain a sunburn, so how was it possible that Bongo's face could redden? The truth was that he felt great shame and embarrassment at the low-life antics of his two companions. Better behaved than any boy could be, Bongo was on his way to becoming a little boy.

Bongo soon taught Tom the nautical art of surfing, for it was better for Tom to be nautical than naughty, but it was no use. He would ride the waves for a while, and then tire of it and look for some poor creature to torment. You can be sure these pranks were making more than ripples in the minds of fairyland, for nothing happens there—or here for that matter—without their knowing about it.

One activity of the other musketeers which Bongo could join in, however, was the meetings with the old lobster who had been Tom's only playmate before meeting Bertie and Bongo. And a very distinguished lobster he was; for he had live barnacles on his claws, which is a great mark of distinction in lobsterdom, and no more to be bought for money than a good conscience or the Victoria Cross.

Tom had never seen a lobster before; and he was mightily taken with this one; for he thought him the most curious, odd, ridiculous creature he had ever seen; and there he was not far wrong; for all the ingenious men, and all the scientific men, and all the fanciful men, in the world, with all the old German bogy-painters into the bargain, could never invent, if all their wits were boiled into one, anything so curious, and so ridiculous, as a lobster.

He had one claw knobbed and the other jagged; and Tom delighted in watching him hold on to the seaweed with his knobbed claw, while he cut up salads with his jagged one, and then put them into his mouth, after smelling at them, like a monkey. And always the little barnacles threw out their casting nets and swept the water, and came in for their share of whatever there was for dinner.

But Tom was most astonished to see how he fired himself off—snap! like the leap-frogs you make out of a goose's breastbone. Certainly he took the most wonderful shots, and backwards, too. For if he wanted to go in a narrow crack ten yards off, what do you think he did? If he had gone in head foremost, of course he could not have turned round. So he used to turn his tail to it, and lay his long horns, which carry his sixth sense in their tips, straight down his back to guide him, and twist his eyes back till they almost came out of their sockets, and then made ready, present, fire, snap!—and away he went, pop into the hole; and peeped out and twiddled his whiskers, as much as to say, "You couldn't do that."

Tom asked him about water-babies. "Yes," he said. He had seen them often. They were meddlesome little creatures, that went about helping fish and shells that got into scrapes. Well, for his part, he should be ashamed to be helped by little soft creatures that had not even a shell on their backs. He had lived quite long enough in the world to take care of himself.

He was a conceited fellow, the old lobster, and not very civil to Tom; and you will hear how he had to alter his mind before he was done, as conceited people generally have, as we have seen before with Toad.

Bertie and Bongo had heard so much of Mr. Lobster from Tom, for they used to sit in holes in the rocks and chat for hours, that Bertie and Bongo convinced Tom to introduce them. So early one morning when the wind was whipping the surface into white foam, so that things were altogether more pleasant below, they set out to explore the rock crevices in search of the lobster.

At midday they had stopped for a salad lunch of sea-lettuce and sea-cucumber, when off in the distance they heard the sharp crack of his huge claw. Swimming in the direction of the sound, they found him inside a trap of green willows, looking very much ashamed of himself, and twiddling his feelers instead of thumbs.

"What, have you been naughty, and have they put you in jail?" jested Tom, when he saw his old friend.

The indignant lobster turned his back on Tom at such a notion, but realizing help might be at hand he turned and said helplessly, "I can't get out."

"*Why* get in, in the first place, is the question at hand," said Bertie.

"Oh," said Tom, "meet my two new friends Bertie and Bongo."

"My pleasure, indeed," said the lobster, twitching one feeler as a countryman might have tapped a forelock.

Just then a rope began to descend from above, and looking up they saw the bottom

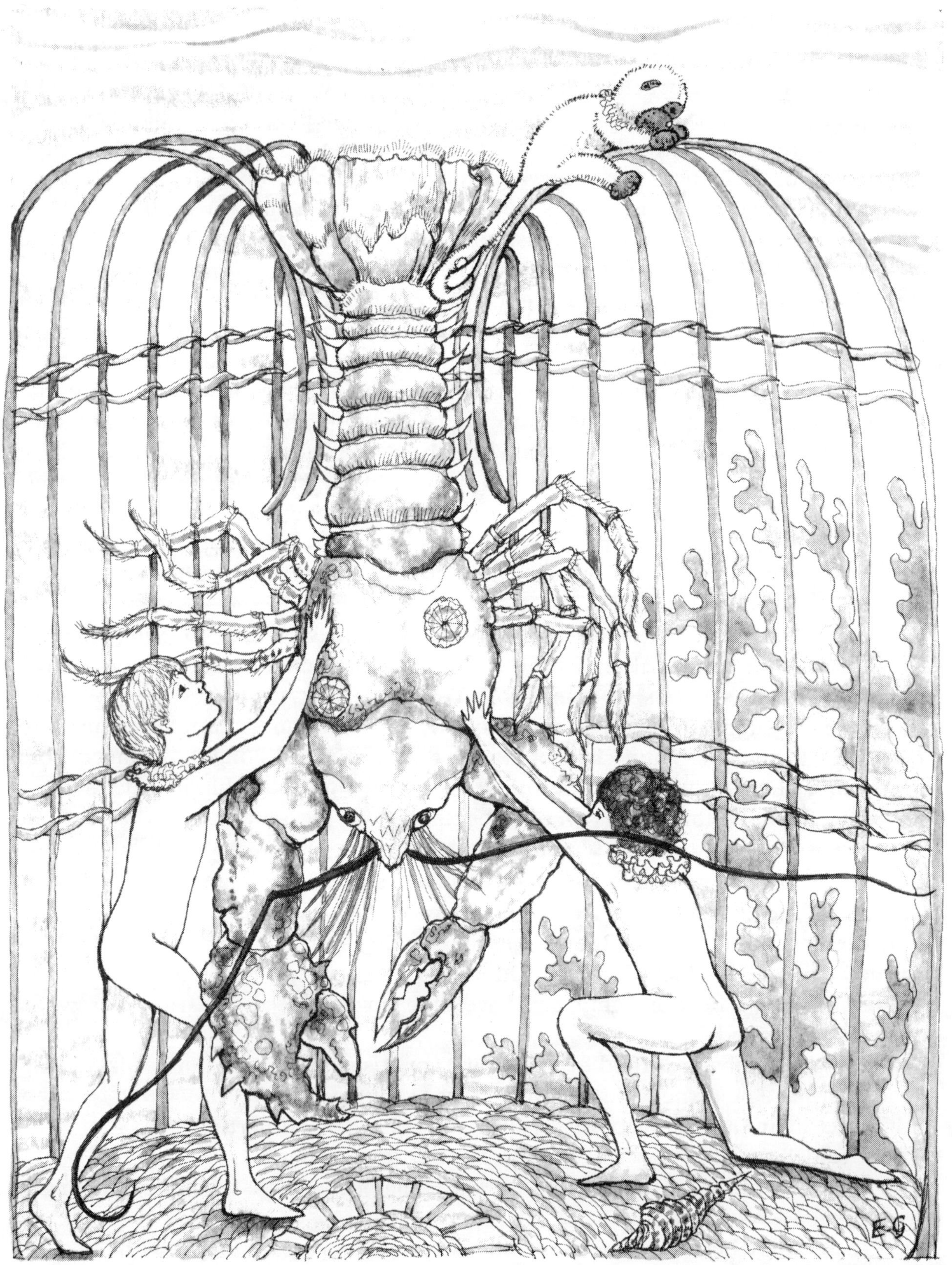
E.G

of a boat. The fisherman had come to haul up his trap!

"We've got to get him out!" cried Tom.

"Where did you come in?!" shouted Bertie.

"Through that round hole at the top."

"Well, then, go *out* through it," said Tom impatiently.

"I can't. Goodness knows I've tried for days and days."

"Well, then, you're for it," said Tom, because just then a hook at the end of the fisherman's rope grappled a branch of the trap, and it began to slowly rise toward the boat.

"One last chance," yelled Bertie. "Here, Bongo, sit on the hole at the top of the trap." Bongo cocked his head and looked at Bertie as if he were crazy, but obeyed him.

"Now Mr. Lobster, back up towards the hole. Bongo see if you can wrap your tail around his flipper and we'll pull on you and he should come out."

But Bertie had not counted on the hard, sharp spikes along the lobster's body, and every time Bongo tried to tighten his tail the spikes made him wince and he had to loosen his grip. Finally the trap was within a few feet of the surface and he knew he would have to hold on despite the pain, and with a mighty heave the lobster was free, and Bongo had two bloody spots on his tail where the lobster's spikes had pierced his skin.

You may guess that this was duly noted in fairyland, for the musketeers had now done a deed befitting a water-baby. Later that same day, Ellie was to make her entrance into fairyland. Oh, so you don't remember Ellie. Have you been paying attention? Well, Tom had not forgotten Ellie I can assure you.

In the afternoon, the surf had died down as the tide went out, and the three musketeers had gone into shore to rest in one of the tidal pools. Not knowing as yet that they like all the water-babies were protected by the fairies, they kept a blanket of algae or seaweed over their heads to hide them from the swooping gulls. From time to time, the musketeer on guard peeped out from under their ceiling of green. Tom was on duty this time and he called out.

"Bertie! There she is!"

"Who?" asked Bertie.

"Why the beautiful young lady I saw sleeping at Harthover House."

"And who is that with her?"

"I do not know, but let's listen."

"And that is a brachiapod," said Professor Ptthmllnsprts, pointing to a shell attached to a rock.

His mother was a Dutchwoman, and his father a Pole, and therefore he was brought up at Petropaulowski; but for all that he was as thorough an Englishman as ever coveted his neighbour's goods.

He was a very great naturalist, and chief professor of Necrobioneopalaeonthydrochthonanthropopithekology in the new university which the king of the Cannibal Islands had founded; and being a member of the Acclimatization Society, he had come here to collect all the nasty things which he could find on the coast of England and turn them loose round the Cannibal Islands, because they had not nasty things enough there to eat.

But he was a very worthy kind good-natured little old gentleman; and very fond of children (for he was not the least a cannibal himself); and very good to all the world as long as it was good to him.

He had met Sir John Harthover somewhere or other, and become very fond of his children. So Ellie and he were walking on the rocks, and he was showing her about one in ten thousand of all the beautiful and curious things which are to be seen there. But little Ellie was not satisfied with them at all. She liked much better to play with live children, or even with dolls, which she could pretend were alive; and at last she said honestly, "I don't care about all these things, because they can't play with me, or talk to me. Now if there were little children in the water, as there used to be, and I could see them, I should like that very much, indeed."

The professor laughed out loud. "Children in the water!" Shaking his head, "You strange little duck."

"I've seen pictures of them," said Ellie.

"Why fanciful pictures aren't scientific proof," scoffed the professor.

"I don't care," said Ellie pouting. "They're so beautiful they *must* be true. The mind would *make* them true!"

"Oh, the mind can't make anything," declared the professor, exasperated with Ellie's persistence. "There are certain *hard* facts which—if you're not careful—one fine day you'll hit your head upon."

"But *why* are there not water-babies?" continued Ellie. "Just because you've never seen one or touched one doesn't mean there can't be any deep, down at the bottom of the sea." And this last she said more with longing than conviction, for she was beginning to cave in to the professor's relentless argument. "Why can't there be?" she said softly once more, almost to herself.

Just then the professor, in his bare feet with his trousers rolled to his knees and a ridiculous sail cap covering his bald head like a tent, trod on the edge of a very sharp mussel and sliced one of his corns quite prettily. He was a scientific man, or so he thought until that moment, but if the fact of the existence of a water-baby were to jump up and bite him, as it was about to, he would still have denied it, because the hurt of his corn and the fall of his dignity made him reply quite sharply to Ellie in a manner of speech that was anything but grammatically logical.

"Because there ain't any water-babies."

Bertie, Bongo, and Tom all broke out laughing at this slip of speech in one who had been so high and mighty in his Latin classifications of all the creatures on the

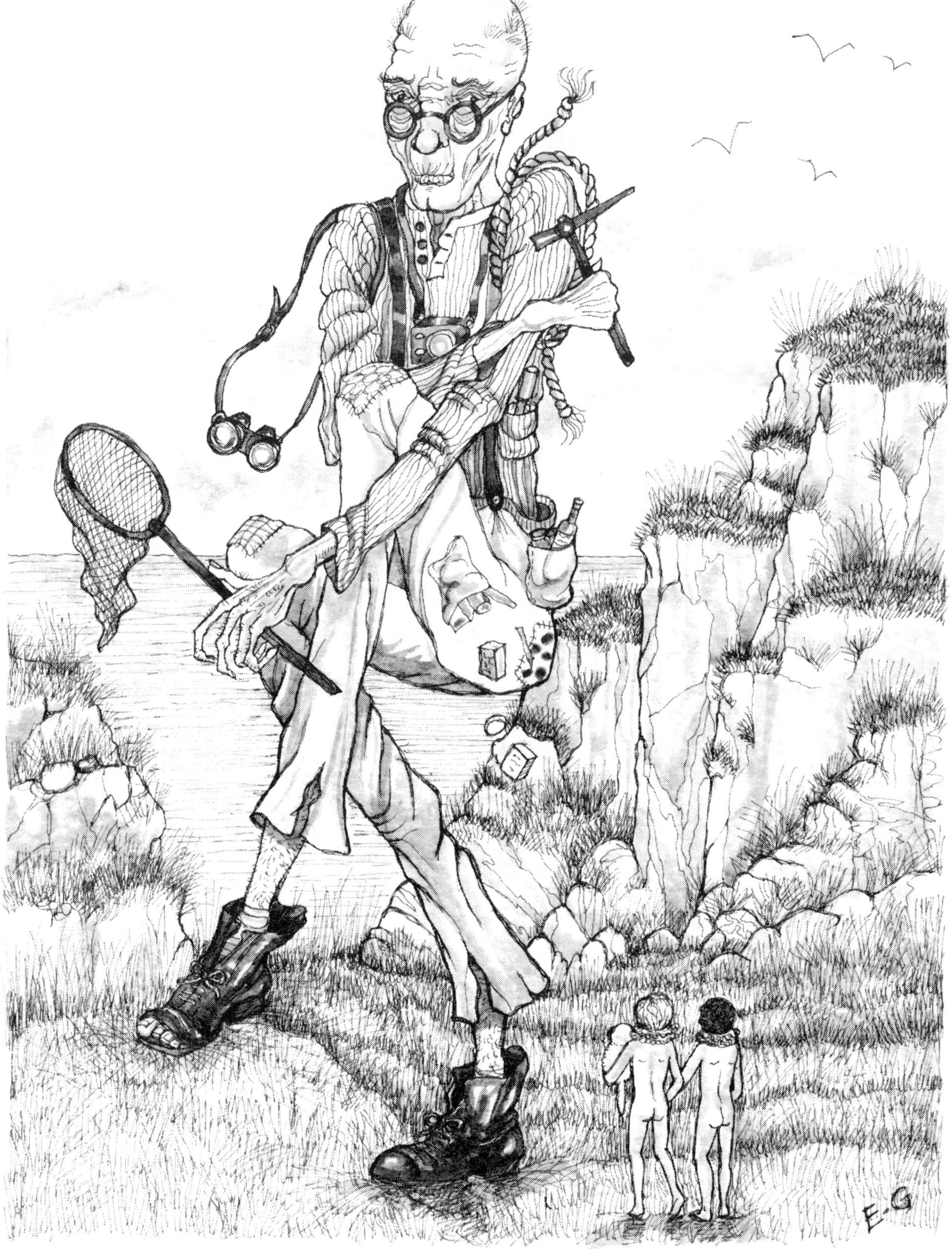
E.G

rock. The professor had hobbled and fallen onto his backside when the indignant Brachiapod (mussel to you) had cut his corn, and he sat practically face to face with Tom, Bertie, and Bongo, in the little pool where they had swam to observe the progress over the rocks of the professor and Ellie. They were pretty well hidden by an undulating piece of seaweed, but the mirth occasioned by his outburst resulted in a stream of bubbles, and being the scientific man that he was, the professor recognized at once the sign of an air-breathing creature and reached in and snatched the seaweed aside to reveal:three tiny creatures resembling humans with a frill of gills on their backs, and one of which had a long tail like a polliwog.

This was Bongo, of course, and right at that moment he wanted very much to be somewhere else, because the professor's hand was poised above them, and more than anything else he feared spending the rest of his life in a zoo, or perhaps, an aquarium. On his part, Bertie was reminding himself to tell Bongo that this was yet another jam he had gotten them into by opening that book back in the library in Hampstead, London, England, oh so long ago.

84

But the professor's hand claimed neither of them, and swooped down on Tom, plucking him by the gills from out of the tidal pool.

"There's a water-baby now," said Ellie, when she saw Tom wriggling between the professor's thumb and forefinger.

"No it ain't! I mean, no it isn't," said the professor, angry at the colloquialism to which Ellie's persistence had brought him.

Tom was in the most horrible fright all the time that the professor held him; for it was fixed in his head that if a man with clothes on caught him, he might put clothes on him too, and make a dirty little chimney-sweep of him again. But when the professor began to squeeze him between his fingers, it was more than he could bear, and between fright and rage, he became as valiant as a cornered mouse and bit the professor's finger till it bled.

"Yah! Yah!" he cried, and dropped Tom onto the seaweed, from where he dived into deeper water, followed closely by Bongo and Bertie.

"It was a water-baby," cried Ellie, "and I want him for my very own." And with that she jumped down off the rock towards the pool where Tom had fallen. But too late! And what was worse, she slipped and fell some six feet, hitting her head on a sharp rock. She lay quite still.

The professor gathered her into his arms, tried to revive her, called to her, and cried over her, but she would not waken at all. She was brought home and put to bed and specialists called in from all about, but no one could help. And after a week, one moonlight night, the fairies came flying in at the window, and brought her such a pretty pair of wings that she could not help putting them on. And she flew with them out of the window, over the land from Harthover House, and over the rocks where she had fallen, and there she paused and flew over and over again, but the fairies did not hurry her, for they never take anyone who is not ready to join them whole-heartedly in fairyland.

She sought Tom, of course. He seemed to her the dearest thing that she had ever seen. Little did she know that he was the same soul that had frightened her in the form of a grimacing black ape back in her room in Harthover House. But now that form, or what was left of it, lay at the bottom of a fast-flowing stream near where it reached the ocean, and Tom's new form cavorted that very moment on the moonlit crests of waves, while Ellie's old form lay in the bed in her room at Harthover House.

The fairies beckoned. She flew one more time over the tidal pool, hoping to see Tom's face shining up at her in the moonlight, but there was only emptiness. She climbed with the fairies towards the stars.

Next day was to dawn eventfully for the three musketeers. They were playing about the white sand on a point of rock when they spied a little figure making a kind of rock garden beneath the waves. It was so intent upon its work that they were quite upon it before being seen. Then the figure cried, "You are not one of us! You must be new babies! Oh, how delightful!" And it hugged and kissed in turn Tom, Bertie, and Bongo. And this was how they found their first water-baby.

At last Tom said, "Where have you been all this time? We have been looking for water-babies for so long."

"There are hundreds of us about these rocks. Did you not hear us each evening when we sing and romp before going home? Now," said the baby, "come and help me, or my work will not be finished when my brothers and sisters come to fetch me home."

"And how shall we help you?" asked Bertie.

"This rock lost all of its pretty flowers in a storm. We must plant it again with seaweeds, coralline, and anemone, and make it the loveliest little rock-garden on all the shore."

And so they worked away at the rock all afternoon, and great fun they had till the tide began to turn. Then Tom heard the rippling, rushing sound of the incoming tide, but beneath that sound came tiny voices laughing like the wind singing, and Tom knew that he had been hearing and seeing the water-babies all along; only he did not know them because his eyes and ears were not opened.

In they came, dozens and dozens of them, some larger, some smaller. And Tom, Bertie, and Bongo too were all welcomed by them with hugs and kisses. Then a ring was made by linking hands, with the three musketeers in the middle, and the water-babies skipped around them, first one way and then another, singing all the time. They made up songs as easily as you and I fall asleep. First there were many songs for Tom's name, because he was now Tom number six, and they danced around him. Then Bertie was put in the middle and they sang that they didn't want his feelings to "hurtie," but they laughed so hard some of them fell down in the sand. Bertie took it all like a good sport, and then suggested that they just try Bert. Then each one in the circle of singing children had a chance to sing a verse that ended in a rhyme with his name. For example:

> "We welcome you beneath the waves, dear Bert.
> And now that you are here, you are sea-girt."

> "We see you travel light.
> And that's all right with us, dear Bert,
> But don't you think you should
> Have brought a single shirt or skirt."

This was a great joke with all of them, of course, because none of them wore a stitch of clothing, and why should they? Were they not babies? When the laughter died down, it became another's turn. A pretty little girl was next to sing, but she was very shy and kept her eyes on the sand as she sang:

"I find you very cute, dear Bert,
Please don't mind me,
If with you I flirt.
And this is how we welcome babies, Bert.
We hope you're not offended by our silly concert."

And then they came to the very last water-baby in the singing and dancing (which they called romping) circle.

"Since you are one of us now, good old Bert,
Don't mind us if we nickname you, Sea-squirt."

This last verse brought the circle down, and they collapsed into an hysterical heap. Bertie stood there alone at the middle of the circle, quite crestfallen. Then the little girl who rhymed his name with "flirt" came up and took his hand and led him back into the circle.

"Don't feel bad," she said. "It's not like public-school up above. No one ever bullies anyone here. And that was your initiation. It wasn't so bad, was it?" Bertie smiled. No, in afterthought, it wasn't bad at all, and already his clever little mind was thinking up name rhymes for the time when a new baby joined them and his chance came.

Well, Bongo's turn was next, and he sat in the middle of the circle with his hands folded like a perfect gentleman. But they had a devil of a time rhyming to his name, and for quite a while this was the best of the lot.

"You're from the far-off Congo,
But with a name like Bongo,
It just won't let my song-go."

Then there were many songs ending in "go" for a time, but the muse threatened to forsake the circle, so the rules were changed to permit rhymes on "Bong," with an "o" added to the end.

"Wherever you may go with us, dear Bongo,
We want you t'know, you can your tail bring along-o."

"Your face is fuzzy, and you hands are mitts, dear Bongo,
But even if we are so different, you belong-o."

And then the song that ended the merrymaking and brought applause for its singer.

"It's true you're just a wee monkey, Bongo,
But in our hearts you're as big as King-Kong-o."

"Now then," they cried all at once, "we must come away home, or the tide will leave us dry. We have mended all the broken seaweed, and put all the rock pools in order, and planted all the shells again in the sand, and nobody will see where the ugly storm swept in last week."

And this is the reason why the rock pools are always so neat and clean; because the water-babies come ashore after every storm, to sweep them out, and comb them down, and put them all to rights again.

Only where men are wasteful and dirty, and let sewers run into the sea, or in any way make a mess upon the clean shore, there the water-babies will not come, sometimes not for hundreds of years (for they cannot abide anything smelly or foul). And that, I suppose, is the reason why there are no water-babies at any watering-place which I have ever seen.

And where is the home of the water-babies? In St. Brandan's fair isle. Did you never hear of the blessed St. Brandan, how he preached to the wild Irish, on the wild wild Kerry coast; he and five other hermits, till they were weary and longed to rest?

So St. Brandan went out to the point of old Dunmore, and looked over the tideway roaring round the Blaskets, at the end of all the world, and away into the ocean, and sighed—"Ah that I had wings as a dove." And far away, before the setting sun, he saw a blue fairy sea, and golden fairy islands, and he said, "Those are the islands of the blest." Then he and his friends got into a coracle, and sailed away and away to the westward, and were never heard of more.

And when St. Brandan and the hermits came to that fairy isle, they found it overgrown with cedars and full of beautiful birds; and he sat down under the cedars, and preached to all the birds in the air. And they liked his sermons so well that they told the fishes in the sea; and they came, and St. Brandan preached to them; and the fishes told the water-babies, who live in the caves under the isle; and they came up by hundreds every Sunday, and St Brandan got quite a neat little Sunday-school. And there he taught the water-babies for a great many hundred years, till his eyes grew too dim to see, and his beard grew so long that he dare not walk for fear of treading on it, and then he might have tumbled down. And at last he and the five hermits fell fast asleep under the cedar shades, and there they sleep unto this day. But the fairies took to the water-babies, and taught them their lessons themselves.

But, on still clear summer evenings, when the sun sinks down into the sea, among the golden cloud-capes and cloud-islands, and locks and friths of azure sky, the sailors fancy that they see, away to westward, St. Brandan's fairy isle.

But whether men can see it or not, St. Brandan's Isle once actually stood there; a great land out in the ocean which has sunk and sunk beneath the waves. Old Plato called it Atlantis, and told strange tales of the wise men who lived therein, and of the wars they fought in the old times. And from that island came strange flowers, which linger still about this land:—The Cornish heath, and Cornish moneywort, and the delicate Venus' hair, and the London-pride which covers the Kerry mountains, and the little pink butterwort of Devon, and the great blue butterwort of Ireland, and the Connemara heath, and the bristle-fern of the Turk waterfall, and many a strange plant more; all fairy tokens left for wise men and good children from off St. Brandan's Isle.

Now when Tom, Bertie, and Bongo arrived there with the other water-babies they marvelled at the sight, for the island stood on vari-colored pillars, beneath which were blue and white grottoes all curtained and festooned with seaweeds, purple and crimson, green and brown, and strewn with white sand, on which the water-babies slept every night.

As the newest arrivals, the three musketeers were given their choice of quarters and chose a purple grotto with black sand, which was situated beneath the main stairway of alabaster marble that descended from the island many fathoms into the sea.

And there were the water-babies in the thousands, more than you could count. All the little children whom the good fairies take because their mothers and fathers will not love them; all who come to grief by ignorance or neglect; all who are overworked or ill-treated, or given drink or drugs when they are young, or let to fall into the fire, or scalded by boiling kettles, or mauled by dogs, or rundown by horses; all the poor children in alleys and tumble-down cottages, who die of famine, cholera, malaria, measles, scarlet fever, and other nasty diseases which one day—pray God—will be no more.

Now it would please me to say that with such pleasant surroundings, and with all his new-found friends, Tom had given up his naughty tricks, and left off tormenting poor dumb animals. But I would not be telling the truth if I said so. True, he had given up shying stones at crabs because Bertie no longer joined him in this sport, and what was the fun in it if there was no one to see you hit the crab or not. But there was one prank he could not forsake, popping pebbles into the anemones' mouths to make them fancy that their dinner was coming.

The other children warned him and said, "Take care what you do, Mrs. Bedonebyasyoudid is coming." But Tom heeded them not, till one Friday morning quite early, Mrs. Bedonebyasyoudid came indeed.

A very tremendous lady she was! All the children stood in a row, very straight indeed, with their hands clasped behind their backs, just as if they were going to be examined by a school inspector.

But although they were very respectful, Tom wanted to make faces at her because she was without a doubt the ugliest person he had ever seen, and it was the custom to abuse such people in the town where he came from. She wore a black bonnet and black shawl in the old style of dress still favored by old women today in the town of Kinsale, County Cork, Ireland. Her nose hooked so much that the bridge started from above her eyebrows and curved down to nearly touch her chin, which

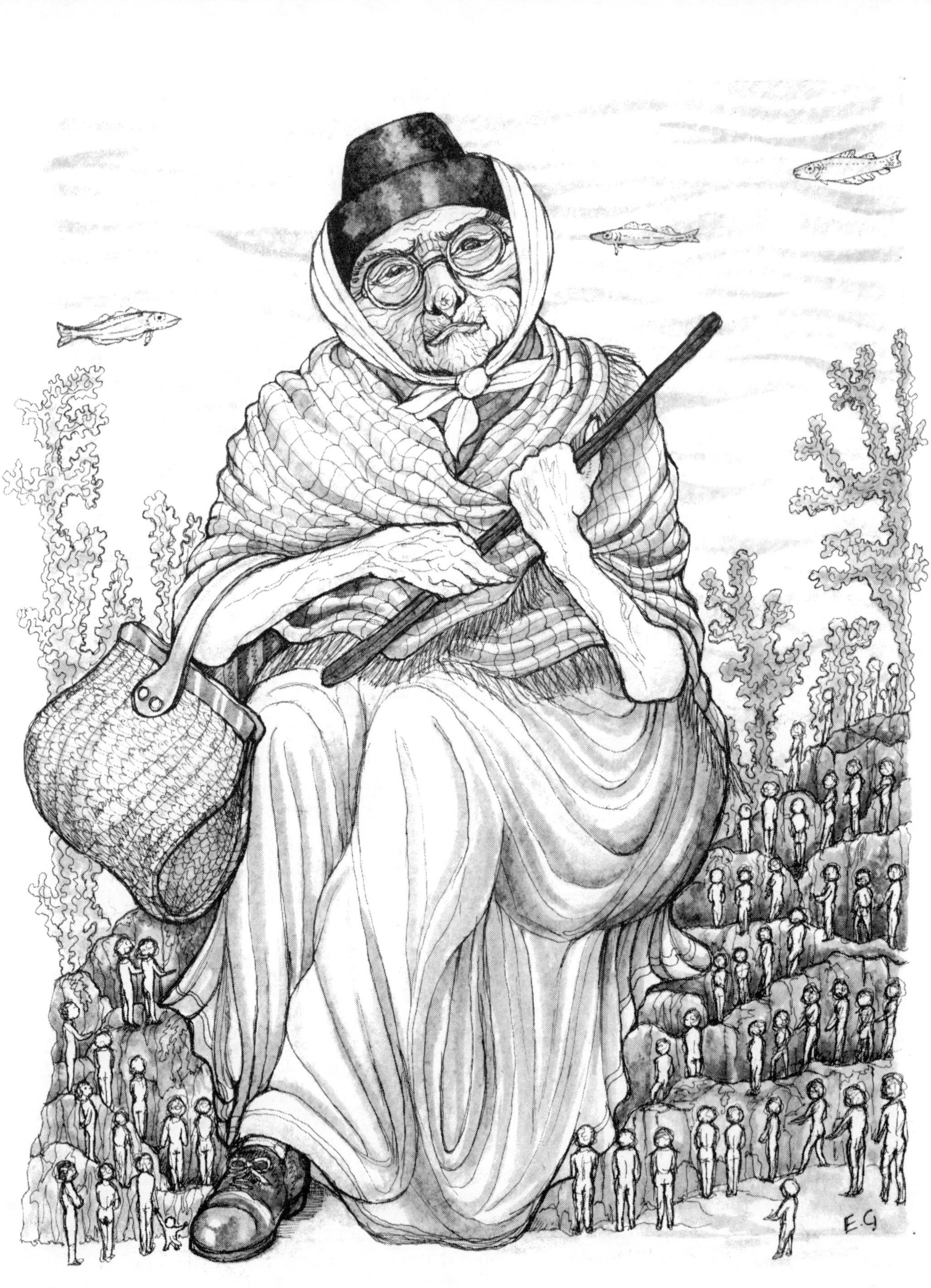

in turn curved wickedly up to meet it. And on that famous nose perched a huge pair of green spectacles. Tom wanted very much to rush at her and pummel her with blows, but the thick birch-rod she carried under her arm dissuaded him. Something about her made him feel very bad indeed.

And when the other children had introduced the new water-babies to her, she gave Tom but a quick glance, while she shook Bertie's hand and smiled at him. Then she spent a minute or so scratching Bongo's head, which pleased him, but darkened Tom's mood considerably.

And she looked at the children one by one, and seemed very much pleased with them, though she never asked them one question about how they were behaving; and then she began giving them all sorts of nice sea-things—sea-cakes, sea-apples, sea-oranges, sea-bullseyes, sea-toffee; and to the very best of all she gave sea-ices, made out of sea-cows cream, which never melt under water.

Now little Tom watched all these sweet things given away, till his mouth watered, and his eyes grew as round as an owl's. For he hoped that his turn would come at last; and so it did. For the lady called him up, and held out her fingers with something in them, and popped it into his mouth; and, lo and behold, it was a nasty cold hard pebble.

"You are a very cruel woman," said he, and began to whimper.

"And you are a very cruel boy; who puts pebbles into the sea anemones' mouths, to take them in, and make them fancy that they had caught a good dinner! As you did to them, so I must do to you."

"Who told you that?" said Tom.

"You did yourself, this very minute."

Tom had never opened his lips; so he was very much taken aback indeed.

"I did not know there was any harm in it," said Tom.

"Then you know now. People continually say that to me: but I tell them, if you don't know that the fire burns, that is no reason that it should not burn you. The lobster did not know that there was any harm in getting into the lobster-pot; but it caught him all the same."

"Dear me," thought Tom, "she knows everything." And so she did, indeed.

"And so, if you do not know that things are wrong, that is no reason why you should not be punished for them; though not as much, my little man" (and the lady looked very kindly, after all), "as if you did know."

"Well, you are a little hard on a poor lad," said Tom.

"Not at all; I am the best friend you ever had in all your life, and I shall go on for ever and ever; for I am as old as Eternity, and yet as young as Time."

And there came over the lady's face a very curious expression—very solemn, and sad; and yet very, very sweet. And she looked up and away, as if she were gazing through the sea, and through the sky, at something far, far off; and as she did so, there came such a quiet, tender, patient, hopeful smile over her face, that Tom thought for the moment that she did not look ugly at all. And no more she did; for she was like a great many people who have not a pretty feature in their faces, and yet are lovely to behold, and draw little children's hearts to them at once; because, though the house is plain enough, yet from the windows a beautiful and good spirit is looking forth.

And Tom smiled in her face, she looked so pleasant for the moment. And the strange fairy smiled too, and said:

"Yes. You thought me very ugly just now, did you not?"

Tom hung his head and got very red about the ears.

"And I am very ugly. I am the ugliest fairy in the world; and I shall be, till people behave themselves as they ought to do. And then I shall grow as handsome as my sister, who is the loveliest fairy in the world, and her name is Mrs. Doasyouwouldbedoneby. So she begins where I end, and I must begin where she ends; and those who will not listen to her, must listen to me, as you will see.

"But you are not all to blame, young Tom, because you had a bad teacher, indeed, but he is very busy now, I can tell you, unlearning his bad ways."

"Then he didn't become a water-baby?" inquired Tom.

"No, not him. And you may have no fear that he'll ever turn you into a sweep again, or beat you until he was too tired to strike again. Perhaps I have been too hard on you, Tom, because you never had the benefit of a teacher or cleric, or mother or father to set your feet on the right path."

And it was true. Working as he did for Grimes from sun up to sun down six days a week, to be in church on the seventh day would have seemed worse than prison, and all Tom knew of God was from hearing Grimes take His name in vain.

"So on Sundays," the fairy continued, "you shied stones at the horses' legs and tormented anything smaller than yourself, just as Grimes had done to you."

Tom blushed from ear to ear.

"Our prisons are full of people like yourself that learn the same kind of early lessons you did, young Tom. So when they start to get bigger, they just can't wait to get their own back—or so they think—on the smallest thing that swims, or crawls, or flies. Did you not start pulling the wings from flies on the very day Grimes discovered he could twist your little arms up your back and make them crack?"

Tom blushed again.

"And after one particularly wicked beating, did you not make your water on an ant-hill so that you could watch them writhe and jump, just as you had done at the hands of Grimes?"

Tom blushed all the way down to his feet.

"Please, mam," said Bertie, "may I ask you a question?"

"You may."

"But what about the people who do all the harm in the first place?" said Bertie. "I know Tom would not have been bad if he had *my* mother and father."

"That's very charitable of you to say that, dear, and I deal with those people on other days. But today is reserved for those who *thought* they were doing good. All the other days are for those who knew that they were doing wrong and got great pleasure from it.

"Now since you three are new babies you shall watch me at my work, and very hard work it is, indeed, for on Fridays I come down here to call up all who have ill-used little children, and serve them as they served the children."

And first she called up all the doctors who give little children too much laxative (they were most of them old ones; for the young ones have learnt better, all but a few army surgeons, who still fancy that a baby's inside is much like a Scottish grenadier's), and she set them in a row; and very rueful they looked, for they knew what was coming.

And first she pulled all their teeth out; and then she bled them all around; and then she dosed them with calomel, and jalap, and salts and senna, and brimstone and treacle; and horrible faces they made; and then she gave them a great emetic of mustard and water, and no basins; and began all over again; and that was the way she spent the morning.

And then she called up a whole troop of foolish ladies, who pinch up their children's waists and toes; and she laced them all up in tight stays, so that they were choked and sick, and their noses grew red, and their hands and feet swelled; and then she crammed their poor feet into the most dreadfully tight boots, and made them all dance, which they did most clumsily indeed; and then she asked them how they liked it; and when they said not at all, she let them go: because they had only done it out of foolish fashion, fancying it was for their children's good, as if wasps' waists and pigs' toes could be pretty, or wholesome, or of any use to anybody.

Then she called all the careless nursery maids, and stuck pins into them all over, and wheeled them about in baby carriages with tight straps across their stomaches and their heads and arms hanging over the side, till they were quite sick and stupid, and would have had sunstrokes; but being under the water, they could have only water-strokes. And mind when you hear a rumbling at the bottom of the sea, sailors will tell you that it is a ground-swell:but now you know better. It is the old lady wheeling the maids about in perambulators.

And by that time she was so tired, she had to go to luncheon.

And after luncheon she set to work again, and called up all the cruel school masters—whole regiments and brigades of them; and, when she saw them, she frowned most terribly, and set to work in earnest, as if the best part of the day's work was to come. More than half of them were nasty, dirty, frowzy, grubby, smelly old monks, who, because they dare not hit a man of their own size, amused themselves with beating little children instead.

And she boxed their ears, and thumped them over the head with rulers, and paddled their hands with canes, and told them that they told stories, and were this and that bad sort of people; and the more they were very indignant, and stood upon their honour, and declared they told the truth, the more she declared they did not, and that they were only telling lies; and at last she birched them all round soundly with her great birch rod, and set them each an imposition of three hundred thousand lines of Hebrew to learn by heart before she came back next Friday. And at that they all cried and howled so, that their breaths came all up through the sea like bubbles out of soda-water; and that is one reason for the bubbles in the sea. There are others: but that is the one which principally concerns little boys. And by that time she was so tired that she was glad to stop; and, indeed, she had done a very good day's work.

"Now I have to be leaving," she said. And looking directly at Tom, "Do as you would be done by, and on Sunday when my sister comes, Madame Doasyouwouldbedoneby, perhaps she will take notice of you, and teach you how to behave."

Well, Tom was determined to be a good boy all day Saturday, and he was. As to Bertie and Bongo, they became very enterprising, and at the entrance to their purple grotto a sign went up, proclaiming "Bongo's Surf Shop." For all their time in the sea, none of the water-babies had discovered the joys of body-surfing, and while Bertie stayed below, booking appointments, Bongo took dozens of water-babies at a time to the waves above. There they were instructed in the proper techniques of this fine art. Before long Bongo was back again for another group. The problem was, once a water-baby started surfing, none of them wanted to stop; and very soon there were more above, riding the waves and frolicking in the sea-mists, than there were below. The sea-gardens and rock pools looked very ragged on that day with no water-babies attending to them. For his part, Bertie had to spend the entire day behind the counter of the surf shop, where the lines of water-babies awaiting Bongo's instruction stretched all the way back to the Irish coast. In just one day, Bongo had become the toast of the fairy isle.

When Sunday morning came, sure enough, Mrs. Doasyouwouldbedoneby came too. All the children began dancing and clapping their hands. Tom, who had never danced before, danced too with all his might, and Bongo showed that he could boogie with the best.

Mrs. Doasyouwouldbedoneby was as different from her sister as day from night. Whereas the latter was of dark complexion, indeed swarthy, with more than hints of a dark beard and moustache, the former was of a light coloring which made her almost too dazzling to gaze upon. Tom found the dark sister the ugliest mortal he had ever seen, and told Bertie as much after she was out of earshot. Bertie replied that he had seen worse, but did not elaborate.

But without a doubt, Tom and Bertie, and all the other children judged the fair sister the fairest in all the world. Unlike her sister, who was gnarly and scaly, she had the nicest soft, smooth, milky skin, and her words fell about the ear like tinkling chimes, whereas her sister's voice was raspy and as grating as a fingernail drawn slowly across a chalkboard.

But it was the manner of the new fairy which made her so attractive to the water-babies. She had the merriest eyes and the sweetest smile. As soon as you saw her, your feet felt like dancing, and your mouth like laughing. So when the children saw her, they naturally all caught hold of her, and pulled at her arms wanting to be picked up, until she sat down on a large rock which served like a throne for her. Then all the water-babies began climbing into her lap, and swinging on her arms, and clinging round her neck, and purring and cuddling like so many kittens. While those who found no room to climb up, fondled the folds of her dress while popping a thumb into their mouths, or sat on the sand and played with her bare toes.

And Tom, Bertie, and Bongo sat staring at her from behind a coral column, for never before had they seen anyone as beautiful.

"And who are you, you little darlings?" she asked.

"Oh, those are the new babies," said the children pulling their thumbs out just in time to answer her question. "They're orphans and never had any mothers or fathers." And with that they all popped their thumbs back in place.

"I did too have a mother and father," said Bertie resentfully.

"Well, now you have me," she said, "and since you are *new* babies you shall have the place of honor."

And she took up two great armfuls of babies—nine hundred under one arm, and thirteen hundred under the other, and tossed them gently, right and left, into the water. But they did not mind and did not even take their thumbs out of their mouths, but came paddling and wriggling back to her like so many tadpoles, till you could see nothing of her from head to foot for the swarm of water-babies.

And Tom, Bertie, and Bongo she raised up, laying Tom aside of her left cheek, and Bertie her right, while Bongo she nestled directly under her chin, and she sang to them the sweetest songs, as she had done to all the children when they had first come to fairyland, and they looked into her eyes and loved her, till they fell fast asleep from pure love.

When they awoke, she said, "Now will you be good boys for my sake and torment no more little creatures till I come again?"

"And then will you cuddle us again?" said Tom and Bertie.

"Of course I will. I should like to take you with me, only I must not. And Bongo, will you take the children to the rock gardens tomorrow, and tidy them up for me?"

Bongo nodded. And away she went.

The very next day, Bongo could be seen sitting in the rock pools intent upon his work, while the other children followed his example.

Now I come to the sad part of my story. You may fancy that Tom was quite good when he had everything that he could want or wish, but you would be very much mistaken. You see, Tom fancied the sea-lollipops, and since he had nothing quite so wonderful before, he began to reason—falsely, of course—that there were quite a few owed to him for all the days in his life before when he had gone without.

He had many a long talk about this with Bertie, who tried to convince him that it was not logical to have as his share more of the sea-lollipops than Mrs. Bedonebyasyoudid allotted to him. Bertie even gave up his own share and that of Bongo's to Tom, so that Tom had thrice his due, but it was no good.

These goodies had now become an obsession to Tom, so that his foolish little head could think of nothing else. He thought of lollipops by day, and dreamed of nothing else at night—and what happened then?!

Then he began to watch the lady to see where she kept the sweet things; and began hiding, sneaking, and following her about, pretending to be looking the other way, or going after something else, till he discovered that she kept them in a beautiful mother-of-pearl chest, away in a deep crack of the rocks. And one night, when all

97

the other children were asleep, and he could not sleep for thinking of lollipops, he crept away among the rocks, and came to the shining chest, and behold! it was open.

But, when he saw all the nice things inside, instead of being delighted, he was quite frightened, and wished he had never come here. And then he would only touch them, and he did; and then he would only taste one, and he did; and then he would only eat one, and he did; and then he would only eat two, and then three, and so on. Then he was terrified lest she should come and catch him, and began gobbling them down so fast that he did not even taste them, or have any pleasure from them. Next he felt sick, and would have only one more; and then only one more again; and so on till he had eaten them all up.

And all the while, close behind him, stood Mrs. Bedonebyasyoudid; and all the while he ate, her features changed. He ate one lollipop; her nose hooked over. He ate another; her chin arched up towards the nose. A third; and her eyebrows beetled out. A fourth; a hairy mole protruded from the chin. And when the last lollipop disappeared down his greedy gullet—you'd have thought Tom was a whale gorging on plankton!—simultaneously all her teeth and hair fell out. Such a sorry sight she was I would not have the artist draw her here for fear of frightening you.

Did she fly at Tom, catch him by the scruff of the neck, hold him, howk him, hump him, hurry him, hit him, poke him, pull him, pinch him, pound him, put him in the corner, shake him, slap him, set him on a cold stone to reconsider himself, and so forth?

Not a bit. For if she had, she knew quite well, Tom would have fought, and kicked, and bit, and said bad words, and turned again that moment into a naughty little heathen chimney-sweep, with his hand against every man, and every man's hand against him.

So she said nothing at all, not even the next day when Tom lined up with all the rest for sweet things. He was horribly afraid of coming—to be sure!—but still more afraid of staying away, lest she suspect him. He was dreadful afraid, too, lest there should be no sweets—he having eaten them all—and lest then the fairy should inquire who had taken them. But, behold! she pulled out just as many as before, which astonished Tom, and frightened him still more.

And, when the fairy looked him full in the face, he shook from head to foot; however, she gave him his share like all the rest. But when he put the sweets in his mouth, he hated the taste of them; and they made him so sick, that he had to get away as fast as he could; and terribly sick he was, and very cross and unhappy, all the week after.

And on Sunday, when Mrs. Doasyouwouldbedoneby came, he wanted to be cuddled like the rest, just as she had done before; but she said very seriously, "I should like to cuddle you; but I cannot, you are so sharp and prickly."

And Tom looked down at himself, and sure enough, he was all over prickles, just like a sea-urchin. Now before he had been a street urchin, and black from head to foot, but now he was a sea-urchin, and more spiny than old Mr. Lobster, and no one in the world could get close enough to cuddle him.

What could Tom do now, but to go away and hide in a corner, and cry? For nobody could play with him now, and he knew full well why.

All that week he was so miserable that, when the ugly fairy came, and looked him once more full in the face, more seriously and sadly then ever, he could stand it no more, and thrust the sweet things aside, saying, "No, I can't bear them now," and then burst out crying, poor little man, and told Mrs. Bedonebyasyoudid every word as it happened.

He was horribly frightened when he had done so; for he expected her to punish him severely. But instead, she raised him up and kissed him, which was not so pleasant, her chin being more bristly than Grimes'; but he was so lonely-hearted he felt rough kissing was better than none.

"I will forgive you, little man," she said. "I always forgive everyone the moment they tell me the truth of their own accord."

"Then you will take away all these nasty prickles?"

"That is a very different matter. You put them there yourself, and only you can take them away."

"But how can I do that?" asked Tom, crying anew.

"Well, I think it is time for you to go to school; so I shall fetch for you a schoolmistress, who will teach you how to get rid of your prickles." And so she went away.

Tom was frightened at the notion of a schoolmistress; for he thought she would certainly come with a birch-rod or cane. But when the fairy brought her, she was the most beautiful little girl that ever was seen, with long curls floating behind her like a golden cloud.

She seemed not to know how to begin, and perhaps would never have begun at all, if poor Tom had not burst out crying, and begged her to teach him how to be good, and help him to cure his prickles; and at that she began teaching him as prettily as ever child was taught in the world.

And what did the little girl teach Tom? That we cannot tell you, because the lessons *under* the sea in fairyland are quite different from those above the sea and on land. But I can tell you this. Much of what we see in the world is illusion, and not really there, although it *appears* to be. Whereas fairyland is thought *not* to be, but is real enough because the mind can go there; and we are here now, as you know from our story.

Therefore, the lessons of fairyland have little to do with learning facts, and much to do with learning truth, for truth can go with you throughout life and in the afterlife as well. And, secondly, like the Greeks of old and the wise men of Atlantis, the lessons of fairy-tales taught one "to know thyself," for if the knower does not know that at least, how can he or she know anything else?

And so the little girl taught Tom every day in the week except Sundays, when she went away and the beautiful fairy took her place. And before too long, his prickles had vanished quite away, and his skin was smooth and clean again like a new-born babe's.

"Dear me!" said the little girl, "now I know who you are. You are the very same chimney-sweep who came into my bedroom back when I lived in the world at Harthover Hall."

"I know you, too," said Tom. "You are the little lady I frightened, and then everyone began chasing me."

And they began telling each other their stories, and Ellie soon realized that Tom was the very water-baby she had seen the professor pluck from the tide pool.

"Oh, you gave him such a bite," laughed Ellie. "He kept arguing with me that there were no water-babies, and I told him he would not know one if it bit him, and so you did," and they both laughed together.

Then, as the weeks passed, Tom felt he would be perfect if he could have but one wish, and that was to go with Ellie, wherever she went on Sundays, for he had begun to love her very much. Ellie would tell him only that it was a very beautiful place, worth all the rest of the world put together. This made Tom long all the more, until one day he said, "Miss Ellie, I will know why I cannot go with you on Sundays, or I shall have no peace, and give you none either."

All that she said was, "Ask the fairies that."

So when Mrs. Bedonebyasyoudid came next, Tom asked her.

"Little boys who are only fit to play with sea-beasts cannot go there," she said. "Those who go there must go first where they do not like to go, and do what they do not like to do, and help someone they do not like."

"Why?!" sulked Tom. "Did Ellie have to do that?"

"Ask her."

Ellie blushed and said, "I did not like coming here at first. I was afraid of you because..."And she blushed again.

"Because I was prickles all over," said Tom. "But I am not prickly now, am I, Miss Ellie?"

"No," said Ellie. "I like you very much now, and I like coming here too."

"And perhaps," said the fairy, "you will learn to like going where you would not go, and helping someone whom you would not help."

But Tom put his finger in his mouth, and hung his head down, for he did not like the idea at all. So when Mrs. Doasyouwouldbedoneby came on Sunday, Tom thought that she was not as strict as her sister, and that he would get an easier answer. Instead, her reply was just the same.

When Ellie came to teach him on Monday he was in a very black mood, indeed, and his discontent with everything round him was such that he did not care to stay.

"I am so miserable here, I'll go; if only you will go with me."

"Ah!" said Ellie. "I wish I might, but the fairy says you must go alone if you go at all. Now don't poke that poor crab, Tom" (for he was feeling very naughty and mischievous), "or the fairy will have to punish you."

Tom was very near saying, "I don't care if she does," but he stopped himself in time.

"I know what she wants me to do," he said, whining most dolefully. "She wants me to go after that nasty old Grimes. But if I find him, I know he will turn me into a chimney-sweep again. That's what I've been afraid of all along."

"Nobody can turn water-babies into sweeps, or hurt them at all, as long as they are good."

"Ah," said naughty Tom, "I see why you have been persuading me all along to go, because you are tired of me, and want to get rid of me."

Little Ellie opened her eyes very wide at that, and they were all brimming over with tears.

"Oh, Tom, Tom!" she said, very mournfully and then she cried, "Oh, Tom! Where are you?!"

And Tom cried, "Oh, Ellie, where are you?"

For neither of them could see each other. Little Ellie vanished away, and Tom heard her voice calling him growing fainter and fainter, till all was silent.

Now when Ellie vanished and Tom began to call her name, it frightened Bertie and he knew not why, but he began crying and screaming like Tom until their voices echoed through all the undersea caverns where all the water-babies heard them and came to see what all the racket was about.

The truth we know is that despite all Tom's blusterings, Tom's cries were those of a young boy who has lost his mother. Tom's feelings for poor Ellie, or perhaps lucky Ellie, carried all the love that he would have given to his mother, had he not been an orphan, and the thought— false as it was—that Ellie might want to be rid of him was more than he could bear. So when Ellie disappeared at Mrs. B's bidding, for it was she who had sent Ellie away, to Bertie it was all too much like his mother's vanishing in the sudden smoke and explosion of the buzz bomb, and Tom's cries caused in him a similar kind of motherless wailing like one tuning fork setting off another in harmonic response.

My what a noise they made! The crabs tunnelled into the sand and put their claws over their ears. All the schools of fish were let out for the day because of the tumult. And the sea snakes got themselves into such an awful tangle that they never could untangle, and even today that part of the ocean where they go round and round chasing their tails endlessly is called the Sargasso Sea.

As a rule, water-babies never cried, so you would think that under the sea was normally a quiet place, and you would be quite right, so you can imagine what a commotion there was under the ocean with *two* boys crying for their mothers all at once. Well, something had to be done, and it would be done by Mrs. Bedonebyasyoudid. Even though she was in another place, she heard and soon appeared. All the water-babies in the sea were gathered in a gigantic circle around Tom, Bertie, and Bongo, who could not cry, because he was stuffed, but felt very much like doing it too.

"You know I only come on Fridays," the fairy said, "so why do you bother me on a Saturday?"

Tom and Bertie both expected her to look very severe, for they were sure she would not take kindly to being put off schedule, busy as she was with putting the whole world right, but she looked fondly at the two boys.

"Well, now, that's over with," said Mrs. B, wiping the tears from their eyes. "And what bitter tears, saltier than the sea."

Tom and Bertie felt like crawling into a sea snake's hole, now that the snakes had all left, for they were the very center of attention with every eye from crab's to waterbabies' upon them.

"The time has come," the fairy said, "to go out now into the world, if ever you intend to be men. Everything in the universe is growing and that is its purpose. Not big in size, but big in spirit, loving and learning, learning and loving, more and more. One day beyond place and beyond time, you both will be as big as I am. But in the lives before you now, Tom must go forward in time and become a man.

"Bertie," said Mrs. B., fixing a compassionate gaze upon him, "you can go a little ways along with Tom, but not far, for you don't belong in this story. Now I have sent Ellie away, and she will not come back for a long time," continued the fairy.

And at that Tom cried so bitterly that the salt sea swelled with his tears, and the tide was nearly half an inch higher than the day before:but perhaps that was due to the waxing of the moon.

"How cruel of you to send Ellie away!" sobbed Tom. "However, I will find her again, if I have to go to the world's end to look for her."

"That's a brave lad," said the fairy, "and just what I wanted to hear. Now you are ready for a Very Great Lesson, which Bertie and Bongo must hear as well. The rest of you water-babies may leave now," and she clapped her hands together and they scattered like a school of tadpoles.

And so, she sat down on a rock, and invited the boys up into her lap, but they shook their heads and stayed at her feet because they were afraid of being scratched by the sharp hairs on her chin.

"As you like," she said, "but you don't mind cuddling on Sundays, do you?"

They knew not the meaning of her words, for cuddling with her sister was quite another matter.

"No, Tom, I shall begin the Very Great Lesson with you. Do you remember on the morning that you set out with Grimes for Harthover Hall, someone joined you on the path?"

"Indeed, it was a tall, handsome Irishwoman," said Tom.

"Now before I disappeared, what did I say to you and Grimes?"

And at this Tom snickered, for Mrs. Bedonebyasyoudid was as different from the Irishwoman as from her beautiful sister, so Tom snickered again at the thought of such an ugly woman comparing herself to the handsome Irishwoman. But then he stopped snickering because he remembered the way she had disappeared, and the way that Ellie was made to disappear, and he wondered if this was one of the powers of fairies.

"You told us about that," said Bertie to Tom.

"Oh," said the fairy, "and did it make an impression upon *you*?"

"Yes," said Bertie, "but I don't know why."

"What were the words then?" she asked, and both Tom and Bertie spoke as one.

"Those that wish to be clean, clean they will be, and those that wish to be foul, foul they will be."

"And so I did," she said. And at that in the twinkling of an eye, the Irishwoman sat before them.

"Now we have got the grime out of you, Thomas," she said, "but not the Grimes. Dirt is dirt, and evil is evil, and the Very Great Lesson is that we are *all* black *and* white, good *and* evil, dirty *and* clean, ugly *and* beautiful."

And with that she changed back into the ugly fairy, Mrs. Bedonebyasyoudid, and never before had she appeared so repulsive. Indeed, Tom and Bertie had to take their eyes from her. And, then, in a trice, she became Mrs. Doasyoubedoneby, and never was she more appealing, and Tom, Bertie, and Bongo clambered into her rosy arms to be cuddled.

"Now, boys, have you got the lesson?"

They had not, and shook their heads.

"The ugliness that you see in me is the evil that you see in yourselves, and that is in the world," she said. "I am like a mirror, reflecting what is there. There is goodness and love too, but perhaps more of evil and hate. It all changes from day to day, and that is why I have to work so hard.

"Fear and hate still bind you to Grimes, Tom. Fear of being turned into a grimy little sweep again. Hate of Grimes for all the suffering he inflicted upon you.

"When Grimes held you, and would not let you be clean, he was demonstrating the main quality of evil:the desire to have power over others, to control them, to bend them to one's own will.

"Grimes was killed poaching another man's salmon. Evil was returned with evil. Evil says you *must* return whatever evil is done to yourself, injury, injustice, slight—call it what you will—with the like amount of evil.

"Now evil is darkness because no choices are possible. Evil begets evil. In the Light, which is the essence of the Good, you have vistas of choices, paths in every direction which you may follow of your own *free* will. Good permits the will of others to work *freely*. Good permits choices, even an evil choice. This is the nature of the Light, the divine principle throughout the universe. But when you are bound in the service of Evil, and make it your master, as you were bound to Grimes, you have no choice but to return evil with evil.

"You learned your lesson only from Grimes, so it is no wonder you had no model for goodness in your life. But when you resisted him—at the spring—your will to be clean rebelled against his power over you. And later, your overwhelming desire to be clean made you slip into the stream, and that is when we took you.

"Now the time has come to confront the evil in yourself that enables Grimes to still have power over you."

Well, at this stage in his young life, Tom thought himself to be a finer lad than ever before, so he was not exactly sure to what evil in him the fairy referred.

"Now I will explain," she said, reading his thought.

"Grimes was killed for poaching another man's salmon. Evil was rewarded with evil. My dark, ugly side, known to you as Mrs. Bedonebyasyoudid, returns ill treatment with ill treatment in the same way. But when she gave you candy, Tom, and you *poached* all her candy, then you were exercising your free will and returning good with evil.

"Now the time has come for you to confront the evil within by returning evil with good. And that single act will place you in the service of the Good, and enable you to go to the deserving place where Ellie goes on Sundays. And that is why we told you, you must *go* where you do not like, *do* what you do not like, and help somebody you do not like. You must return Grimes' evil to you with Good, or fear and hate will bind you to him forever. Now Tom, class is dismissed for you, and you may join your playmates. As for Bertie and Bongo, the lesson begins in earnest for you now."

So Tom reluctantly slid down from her lap and went off in search of the other water-babies.

Bongo and Bertie snuggled in closer to her, and could not remember a happier time. Having her all to themselves was paradise. But she soon disrupted their tranquil mood.

"Now Bertie," she said, "what about your Purpose?"

Strange, it was so important to him once upon a time, and now he had nearly forgotten.

"Well, my purpose was to find my mother," he said, "but now I have you, and you're as wonderful as any mother could be!"

She smiled warmly at him, but said, "I am everyone's mother. In time everyone comes to me. But your time is not yet because fairyland did not take you, but you came to fairyland, therefore, you cannot stay."

At the thought of losing so much warmth and love, comfort and caresses, Bertie burst into tears. Bongo cocked his head and looked up at the lovely lady to inquire if Bertie's going meant that he would have to go too. She nodded, and said, "What would Bertie do without you?" Bongo, of course, could not cry because he was stuffed, but he balled his hands into fists and rubbed them around in his eyes as if he were actually crying. The good fairy began scratching Bongo's head very slowly and thoughtfully, and very soon he began to feel better.

"Your purpose," she said comfortingly to Bertie, "is tied to a greater Purpose in the universe. You are going to be one to show the world new models of space and time, and this will help to free the world's thinking from being bound to Matter. There is also Light, its opposite.

"You will have to grow up in fairyland, and become a man in mind, if not in body, because no one who goes back into the world from fairyland changes in form. You will have help from me, and others, for you are protected—as you already know—as are all creatures in the Light.

"In order to return to the world to confront your destiny, you must solve the Mystery of Space and Time. But the Mystery of Space and Time which you seek is unfathomable. In fairyland you are outside of space and time and must get back into them."

"Then how can I solve it?" said Bertie.

"I will give you a clue," she said. "Space is not space, and time is not time. What appears to be these things to you is, indeed, unfathomable, because it is an illusion, without true reality. When you discover the *true* nature of space and time, then you can begin to direct your mind to the solution of the Mystery. But solving an illusion would not be a solution, and would not land you in London again, but heaven knows where.

"Now I must examine your soul as I did Tom's, for where a kernel of evil lurks there, you cannot succeed in your purpose. Now answer straightaway without hesitation. Whom do you hate?"

Bertie thought quickly, and said, "No one."

And then, in barely a moment, he found himself in the lap of the ugly fairy, and she looked very stern, indeed.

"Aren't you forgetting why you are an orphan?" she said, but her voice was gentle with no hint of a reprimand.

Then Bertie thought of the great mechanical bird that had descended on his house and had taken his mother and father from him, and he began to cry.

"There is a special place for the man who built that bird," she said, reading his mind. "He has done the same evil to millions of people. He will be at his lessons for many a day to forever. Time will be fairly run down by the time he finishes, for the undoing takes much longer than the doing."

Bertie thought of Hitler suffering at the hands of Mrs. Bedonebyasyoudid, and it made him feel much better, and he blew his nose and stopped crying.

One thing, however, puzzled him in all that the fairy lady had said. "Please, mam, may I ask you one question?"

"To be sure, little man."

"Tell me this, if you please. How can Hitler be alive and still doing mischief in the world, and already in *your* place?"

"One's deeds go before one. What you do in life prepares the place that you go to, so you are *there* already working for Light or Darkness. Remember, working for Light *frees* you; working for Darkness binds you. Remember that if you remember nothing else of the lessons of fairyland.

"Now the time has come to start you on your adventure. Tom will accompany you part of the way, since you are, after all, *three* musketeers."

"I'll appear, when you need me, to show you the way out of the story. Your purpose, as you know in your heart of hearts is not forward in time, but backwards in time. That is very difficult for humans to accomplish, and even for fairies too, since the timing must be perfectly right. But I will help you as much as I am able."

In a trice, Mrs. B. summoned Tom to her once again. "Now Tom take your thumb out of your mouth and listen to me. What Grimes has to do with Ellie is *your* becoming a man," said the fairy, "and if Ellie is your bliss, you must follow her.

"Bertie goes to the end of your story with you because he too has lessons to learn on the path to his purpose.

"And Bongo goes because while his two friends were intent on becoming little monkeys, he has forsaken the mischievous ways of monkeys, and wishes with all his heart to become a little boy."

Her voice now spoke sternly to them, and they cast down their eyes as she said, "I must say, so far in this story, little Bongo has shown himself to be far more a gentleman than either of you two. Stones in anemones' mouths, indeed! And stealing my sea lollipops!"

"We were just playing," said Tom.

"What?" Mrs. B. said sharply.

"Playing at being bad," said Tom, "so that we could show you how good we can be."

"Well I never," Mrs. B. replied. "That answer takes the cake, and all the icing too."

"Sure," said Bertie. "Why if I hadn't met Tom and had some fun, I could have grown up to be an old fuddy-duddy."

"True enough," said Mrs. B. "That's why your purpose took you here. But there's an even greater lesson to be learned, and that is that by playing you see through the illusion of space and time. You create your *own* space and time, your own places and people, and you make them *real* in your mind just as surely as any god ever made a universe. Now I have told you, Bertie, that you will be one to bring Light to the world, but you cannot do that while there is still a kernel of darkness within you, and so you must accompany Tom to where Grimes awaits him."

Tom was about to exclaim, "Do I have too?!" Fortunately for our story he thought better of it. He was now beginning to see that growing up did not involve doing as he pleased, and more often than not doing as he did not please. The rest of you will find that to be true, I am certain, if you decide to grow up in reality as well as in size.

But Mrs. B. was not yet through with her instruction to Bertie.

"Now, Bertie, I'll speak frankly, and not mince my words. Your personal purpose is to somehow save your mother and father if you can get back into the same moment in time when they were killed by the flying bomb. That is good, and shows you are in the service of the Light, because you are thinking of others, and have no selfish goal for yourself.

"In Fairyland, where you are now, you are *out* of time, neither in the past nor the present, so only by solving the Mystery of space/time will you be able to re-enter the world at the appropriate time and place.

"But unbeknownst to you, all the eyes of Fairyland have been turned upon you ever since your coming because we *take* people, but they are not allowed to come here. In your case, a kind of cosmic exception has been made because your personal purpose is linked inevitably to the Greater Purpose of which I spoke to you earlier. *If* you solve the Mystery of space/time, and somehow get back into the world, your journey will bring Light to minds now bound by Matter."

"I see," said Bertie. "Every moment is an instruction."

"Not that heavy," answered Mrs. B. "Every moment is an en*light*enment, with the accent on the light. Remember that when you are near being overwhelmed by perils and darkness.

"And remember that it is all illusion, a play of light and darkness, shadows that seem real. When you slay a monster, you kill a dark part of yourself. Better to bring it to the light, and transform it into an ally."

"I still don't see why I have to find Grimes," intruded Tom. "He can never do me any good, and he has already done me a lot of harm."

"Precisely why you *must* find him," said Mrs. B.

"Now boys, I've said all I'm going to say, and the time has come to be men of action. But you must go farther than the world's end, if you want to find Mr. Grimes; for he is at the Other-end-of-Nowhere. For a start though go to Shiny Wall."

"Oh, but I do not know the way to Shiny Wall," cried Tom.

"Nor me neither," added Bertie.

And just for good measure, Bongo shook his head too.

"Little boys must take the trouble to find out things for themselves, or they will never grow to be men. Ask all the beasts in the sea, and all the birds in the air, and if you have been good to but one of them, one will tell you the way to Shiny Wall."

And with no more of an *adieu*, Mrs. B. vanished.

"I suppose we'd better start swimming," said Bertie, "unless we want to be babies forever."

And so, not knowing where they were going, but knowing that they had to be going, they set out swimming steadily towards the West, which we know was not at all in the direction of Shiny Wall, or at least, I the author know that, but it is the custom of young heroes when beginning their adventures to follow the direction of the sun as it gallops across the sky from east to west.

And after a time, to set the pace, Bertie began to call out to his comrades, "Stroke! Stroke!" like a coxswain in a rowing shell, and so they swam all that day and the next.

Little did they know that they had set the marathon record for water-baby swimming, but they were spent enough to know that their bodies needed a rest. Finding a convenient rock, the three musketeers hauled out of the sea, and draped themselves like three basking barnacles on the sunny side of the rock.

They were too tired to talk, and anyhow no one wanted to say a word, because they were very disappointed at swimming so far with nary a sign of Shiny Wall. Their discontent would have been more acute had they known that they had been swimming for two days in a complete circle.

But fate would play its hand in their adventure, for at that moment old Mr. Lobster crawled out of the mud, where he had been sitting below, and onto the rock to sun himself in the afternoon rays. What! You did not know that lobsters were given to basking in the sun, even as you and I? The proof is in the red color of the lobster's shell, of course, for how else can he come by that color?

The boys were happy to see an old friend, sure enough, and Tom remarked to him, "You must have been swimming along beneath us all this time!"

"Not I," said the lobster.

"Then how did you get here?" asked Bertie.

"I have been *here* since you left," said the lobster somewhat crustily. "The trap you sprang me from is but a short crawl from here."

And then the musketeers knew that they had been swimming in a circle, and Tom thought of Ellie, whom he would never find, and he began to cry. And Bertie thought of his mother and father, whom he would never find, and he began to cry. And Bongo thought that swimming in circles he would never become a little boy, and he balled up his paws into his eyes and tried very hard to cry.

"Now we'll never find Shiny Wall, and I'll never find Ellie, ever, ever again," wailed Tom.

"Shiny Wall?" said the lobster, perking up his antennae.

"Yes, Shiny Wall!" wailed Bertie.

Now their goal, Shiny Wall, had become an object of vexation because of their great frustration in not being able to find it.

"Well, as you helped me, now can I return the favor," said the lobster quietly amid the hullabaloo.

For a moment, no one reacted, and then both boys stopped crying and looked in the direction of the lobster. When he saw that he had their attention, he spoke again.

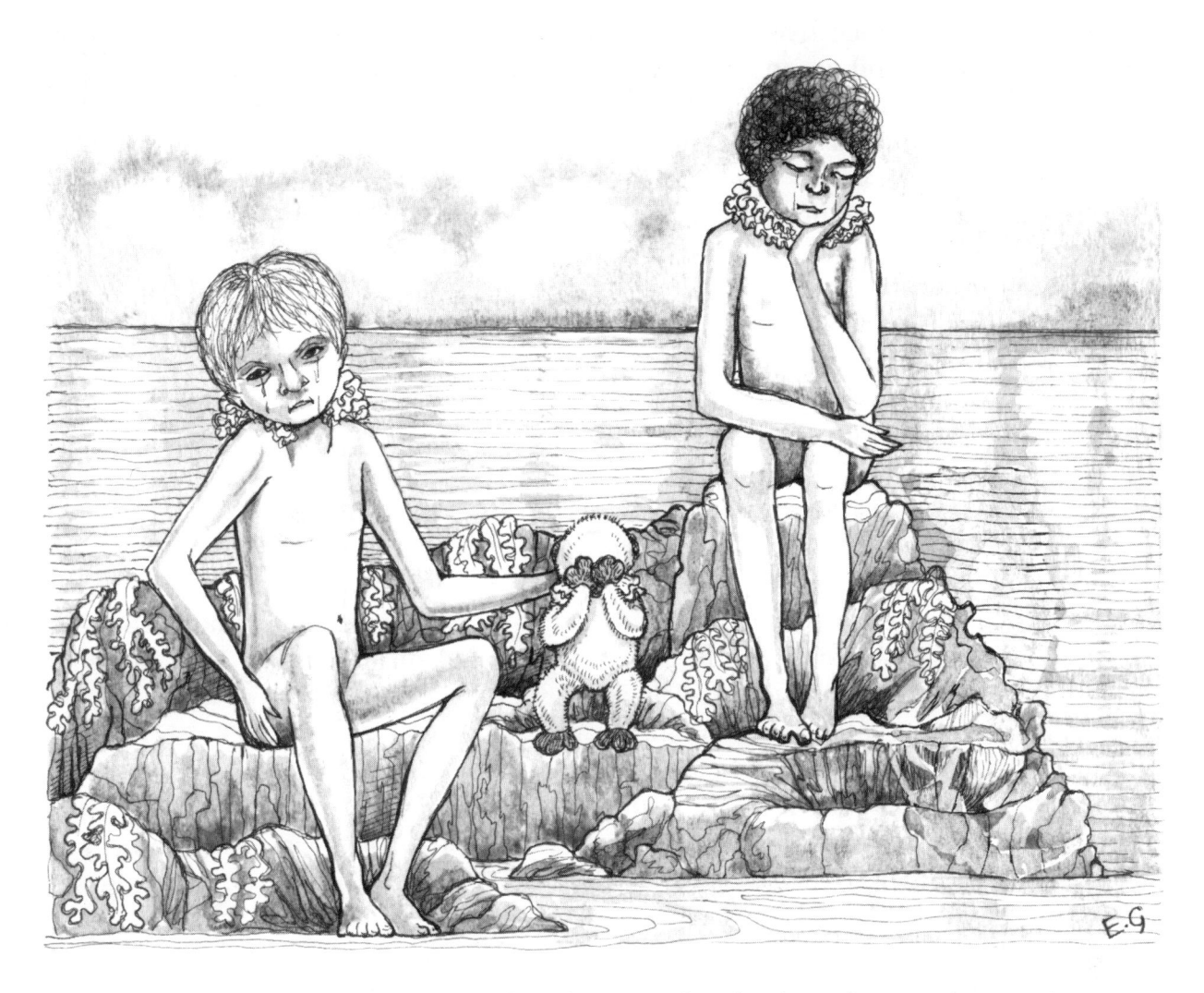

"If I were you, young gentlemen, I should go to the Allalonestone, and ask the last of the Gairfowl. She came from Shiny Wall, or so she told me, many years ago. She is of a very ancient clan, and knows a good deal more than these modern upstarts."

And so they rested that night with old lobster, and set out again next day on their adventure, armed with the instructions he had given them to follow to Shiny Wall. Away they went due north for seven days and seven nights, till they came to a great codbank, the like of which had never been seen before. The great cod lay below them in tens of thousands, and gobbled shell-fish all day long, and the blue sharks roved above and gobbled them when they came up. So they ate, and ate, and ate each other, as they had done since the making of the world; for no man had come here yet to catch them, and find out how rich is old Mother Carey.

And in this place, standing up on the Allalonestone, all alone, they saw the last of the Gairfowl. And a very grand old lady she was, full three feet high, and bolt upright like some old Scottish Highland chieftainess. She had on a black velvet gown, and a white pinner and apron, and a very high bridge to her nose (which is a sure mark of high breeding), and a large pair of white spectacles on it, which made her look rather odd; but it was the ancient fashion of her house.

And instead of wings, she had two little feathery arms, smaller than a penguin's wings, with which she fanned herself, and complained of the dreadful heat.

The three musketeers bowed deeply before her, and then humbly inquired of the way to Shiny Wall.

"Shiny Wall?" she said. "Who should know better than I? We all came from Shiny Wall, thousands of years ago, when there were still many glaciers to ride upon, and the climate was fit for gentlefolk. But now, what with the heat, and these vulgar-winged things who fly up and down and eat everything, one hardly ventures off this rock.

"Once we were spread over all the Northern Isles, a great nation, indeed. But men shot us, and knocked us on the head, and took our eggs, till at last there were none of us left, except on the old isle of Gairfowlskerry, just off the coast of Ireland, which no man could climb. But one day there came a terrible quaking of the earth, and the sea boiled, and the air was filled with smoke and dust, and our land vanished beneath the sea. I am the last of the clan, and here I sit, all alone."

And she began to cry tears of pure oil.

"But, please, which is the way to Shiny Wall?" pleaded Bertie.

"Oh, must you go, must you go?! I am so all alone. Let me see. Really, my poor old brains are getting quite puzzled. Do you know, my little dears, I am afraid you must ask some of these vulgar—how shall I say it—*flying* birds, for I have quite forgotten."

And just as the boys were about to cry again, there came a flock of petrels, who are Mother Carey's own chickens. Much prettier, they thought, than Lady Gairfowl, and perhaps they were right, for it had been a very long time since Mother Carey had invented the Gairfowl and the time that she had invented them. They flitted along like a flock of black swallows, and hopped and skipped from wave to wave, lifting up their little feet behind them so daintily, and whistling to each other so tenderly, that Tom fell in love with them at once, and called to them to know the way to Shiny Wall.

"Shiny Wall? Do you want Shiny Wall? Then come with us, for Mother Carey sends us out over all the seas to show the good birds the way home."

And the three musketeers thought that their chance had at last come, and were all agog to start for Shiny Wall, but the petrels said no they must go first to Allfowlness, and wait there for the great gathering of all the seabirds, before they start for their summer breeding places far away in the Northern isles. There they would surely find some birds who were going to Shiny Wall.

And after a while the birds began to gather at Allfowlness, in thousands and tens of thousands, blackening all the air; swans and brant geese, harlequins and eiders, harelds and gaganeys, smews and goosanders, divers and loons, grebes and dovekies, auks and razor-bills, gannets and petrels, skuas and tern, with gulls beyond all naming or numbering; and they paddled and washed and splashed and combed and brushed themselves, till the shore was white with feathers; and they quacked and clucked and gabbled and chattered and screamed and whooped as they talked over matters with their friends, and settled where they were to go and breed that summer, till you might have heard them from ten miles away.

Then the petrels asked this bird and that whether they would take Tom, Bertie, and Bongo to Shiny Wall:but one flock was going to Sutherland, and one to the Shetlands, and one to Norway, and one to Spitzbergen, and one to Iceland, and one to Greenland; but none would go to Shiny Wall. So the good-natured petrels said they would show them part of the way themselves, but they were only going as far as Jan Mayen's land; and after that they must shift for themselves.

And then all the birds rose up, and streamed away in long black lines, north, and north-east, and north-west, across the bright blue summer sky; and their cry was like ten thousand peals of bells.

As the petrels and the three musketeers went north-eastward, it began to blow right hard, but the petrels never cared, for the gale was right abaft, and away they went over the crests of the billows, as merry as so many flying-fish, with the three musketeers behind them, riding the foam-bedizened crests in the surfing style Bongo had taught them.

At last they saw an ugly sight—the black side of a great ship, water-logged in the trough of the sea. Her funnel and her masts were overboard, and swayed and surged under her lee; her decks were swept clean as a barn floor by the waves, and no living soul was to be seen on board.

The petrels flew up to the floundering ship, and circled the water round her, for they expected to find some floating salt pork, but were sorely disappointed, and so wailed round and round. The musketeers scrambled on board, grateful for a place to rest at last.

And there, in a little cot, lashed under the bulwark, lay a baby fast asleep, and as they went to look at it, from under the cot jumped a little terrier, barking and snapping, and would not let them touch the cot. But they wanted to rescue the baby from the doomed ship, and as they were struggling with the dog, a tall green sea walked in over the weather side of the ship, and swept them—baby, dog, and all—into the sea.

"Oh, the baby, the baby!" screamed Bertie. But in a moment he saw the cot settling down through the green water, with the baby smiling and laughing up at him. Then he saw the two fairy sisters, Mrs. Bedonebyasyoudid and Mrs. Doasyouwouldbedoneby, rise up from below and cradle the baby gently in their soft arms, and he knew that it was all right, and that there would be a new water-baby in the kingdom below.

And what of the poor little dog? Why, after he had kicked and coughed a little, he sneezed so hard that he sneezed himself clean out of his skin, and turned into a water-dog, and jumped and danced round the musketeers, and ran over the crests of the waves, and snapped at the jelly-fish and the mackerel, and followed them all the way to the Other-end-of-Nowhere.

Then they went on again, till they began to see the peak of Jan Mayen's Land, standing up like a white sugar-loaf, two miles above the clouds. And there they fell in with a whole flock of mollymocks, who were feeding on a dead whale.

"These are the fellows to show you the way," said the petrels. "We don't like to get among the ice pack, for fear of nipping our toes; but the mollys dare fly anywhere." So the petrels called to the mollys, but they were so busy fighting over the blubber that they did not take the least notice.

Come, you lazy greedy lubbers," called the petrels to the mollys. "Here are three lads and their dog on their way to Mother Carey, and if you don't attend to them, you won't earn your discharge from her."

"Greedy we are," says a fat old molly, "but lazy we ain't And, as for lubbers, we're no more lubbers than you. Let's have a look at the lads."

And he flapped right into Bongo's face, and stared at him in the most impudent way.

"It's time this fellow saw a razor, for his face is as fuzzy as a new-fallen coconut," said the old mollymock.

Had he been able to, Bongo would have blushed, but he was relieved when the old molly waddled up to Tom.

"Where do ye hail from?" he asked, and then the same of Bertie. He seemed pleased with their answers, and their pluck in coming so far from home.

"Come along, lads," he called to the rest of the flock. "We've eaten enough blubber for one day, and we can work out a bit of our time by helping these lads."

So the mollys took them up on their backs, and flew far up into the sky, until even the carcass of the whale seemed very small below. The poor dog was terribly frightened and barked and barked to show his displeasure. How easily he had adapted to swimming, but this was an entirely different matter.

Bongo thoroughly enjoyed the experience. Pretending to be an aviator, and the molly his plane, he imagined himself to be in a formation of bombers crossing the English channel on their way to Germany.

The birds kept up a good-natured bantering between themselves as they flew.

"Who are you jolly birds?" called Tom.

"We are the spirits of the old Greenland skippers (as every sailor knows), who hunted here, right whales and horse-whales, full hundreds of years ago. But because we were saucy and greedy hunters, we were turned into mollys, to eat dead whale's blubber all the rest of our days."

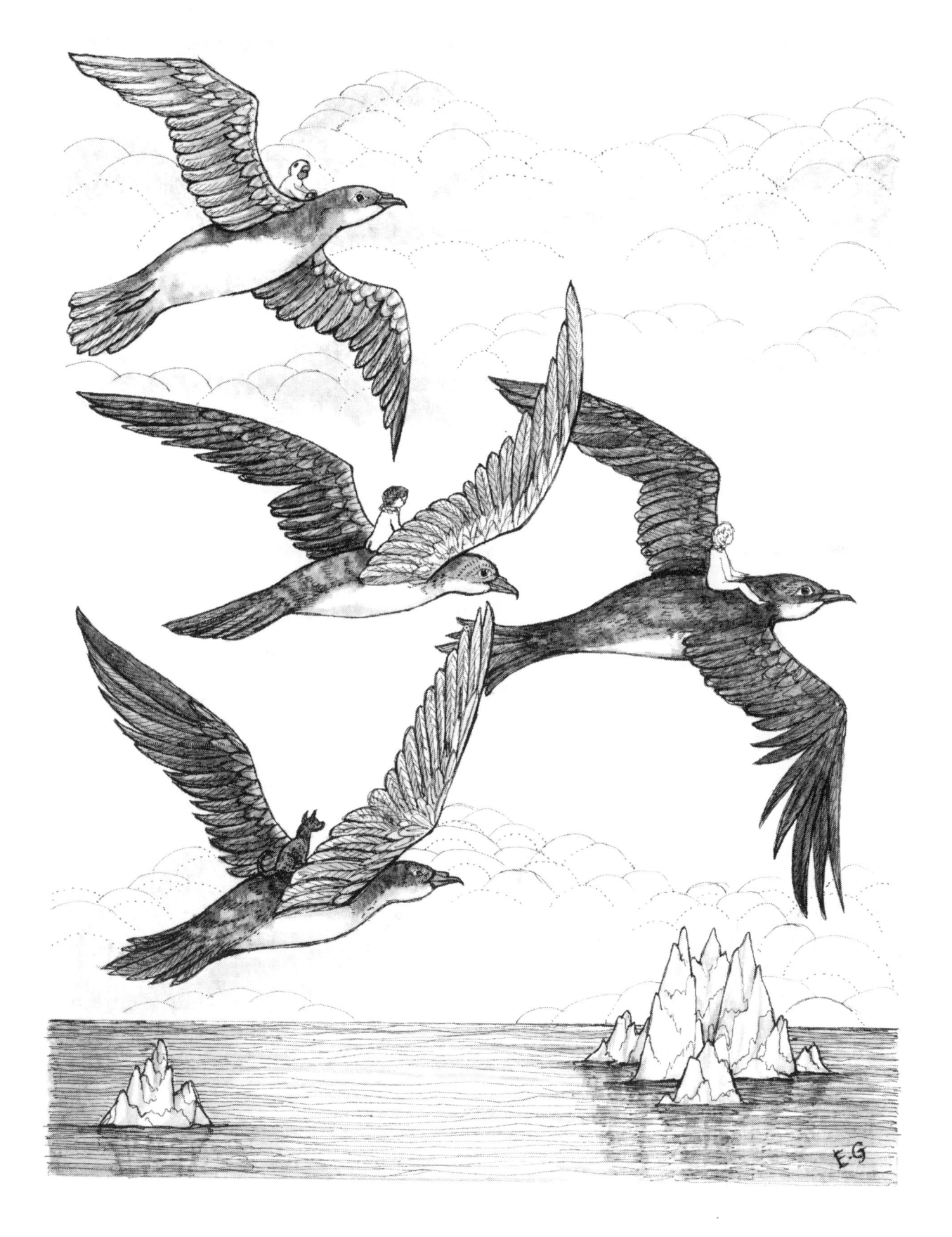
E.G

"And who were you, when you lived long ago," asked Tom of the old molly, for he saw that he was the leader of all the birds.

"My name is Hendrick Hudson, and a right good skipper was I, for I discovered the Hudson River, and I named Hudson Bay, and many have followed in my wake all the world over. But I was a hard man in my time, that's truth, and stole the poor Indians off the coast of Maine, and sold them for slaves down in Virginia. And at last I was so cruel to my sailors, here in these northern seas, that they set me adrift in an open boat, and I was never heard of more. So now I am the king of the mollys, living off dead carcasses as I did once live on whales and slaves, until I've worked out my time with Mother Carey."

Now they approached the edge of a gigantic ice pack, and beyond it they could see Shiny Wall looming, through mist, and snow, and storm. But the pack rolled horribly upon the swell, and ice giants fought and roared, and leapt upon each other's backs, and ground each other to powder, and the three musketeers knew then that had they been swimming there they would have been crushed to death.

And they were more afraid when they saw lying among the ice pack the wrecks of many a gallant ship; some with masts and sail all standing, and the brave seamen standing straight at their posts, frozen fast like statues. But the good mollys soared easily over the roaring ice giants, and set down our friends at the foot of Shiny Wall.

"But where is the gate through it?" asked Tom.

"There is no gate," said the mollys, "and never a crack of one, as better fellows than you lads have found to their cost."

"How are we to go on then?" cried Bertie in despair.

"Dive under, if you have the pluck, but who knows or not if the ice goes all the way to the bottom."

"Well, we've not come this far to be turning around now," said Tom, and the others all agreed. Then bidding farewell to the mollys, they turned turtle and became true water-babies again.

Down, down, down, they went; and the deeper they dove, the darker it became, until the only way they could stay together was by holding hands and paws and kicking with their little legs.

At last they saw light again, a kind of shining that came from the bottom of the huge ring of iceberg which surrounded Mother Carey's Peacepool. Then up they swam, through a thousand fathoms or more, through clouds of sea moths, with pink heads and wings and opal bodies, and jelly-fish of all the colors of the world, that fluttered through the water like flapping parasols. The dog snapped at them till his jaws were tired, and Bongo tried petting them till he found they had a nasty sting.

Then all at once they broke the surface to a dazzling sight. All round the pool—miles across—ice cliffs rose, in walls and spire and battlements, and caves and bridges, in which the ice fairies live, driving away the storms and clouds so that Mother Carey's pool may lie calm from year's end to year's end. And the sun acted as policeman, and walked around outside each day, peeping just over the top of the ice wall to see that all went right; and now and then he played conjuring tricks, or had an exhibition of fireworks, which sometimes shoot so high into the sky that they came to be seen all the way to Scotland, where they are called the Northern Lights. All this the sun did to amuse the ice-fairies. Sometimes he would make himself into four or five suns at once, or paint the sky with rings and crosses and crescents of white fire, and stick himself in the middle of them, and wink at the fairies.

And here in Peacepool the good whales lay, the happy sleepy beasts upon the azure sea. There were right whales, pilots, finners, and razor-backs, and bottle-noses, and spotted unicorns with long ivory horns. But the sperm whales are such raging, ramping, roaring, rambunctious fellows, that if Mother Carey let them in, there would be no peace in Peacepool. So she packs them away in a great pool by themselves at the South Pole, and there they butt each other with their ugly noses day and night.

And so our friends began to swim towards the iceberg in the middle, but when they came near it took the form of the grandest old lady they had ever seen—a white marble lady, sitting on a white marble throne. And from the foot of the throne, there swam away out into the sea, millions and millions of new-born creatures, of more shapes and colors than man ever dreamed. And they were Mother Carey's children, whom she makes out of sea-water all day long. She sat quite still with her chin upon her hand, looking down into the sea with two great grand blue eyes, as blue as the sea itself. Her hair was as white as snow—for she was very old, in fact, as old as anything you are likely to come across except the difference between right and wrong.

When she saw Tom, Bertie, Bongo, and the little dog, she turned her gaze kindly upon them and said, "How can I help you? It is very long since I have seen water-babies here."

"If you please, mam," said Bertie, "we need to find the way to the Other-end-of-Nowhere."

"You ought to know yourselves, for you have all been there before."

They all looked at her with puzzled faces.

"I'm sure we have forgotten then," said Tom.

"Then look at me," she said.

And as they gazed into the depths of her great blue eyes, it was like being born again, and they recollected the way perfectly.

"Thank you, mam," said Tom, finally breaking her gaze. "Then we won't trouble your ladyship any longer, for I hear you are always very busy making new beasts out of old."

"So people fancy," she said. "But I sit here and watch them make themselves."

"How is that possible?" asked Bertie.

"Know, silly child," she said, "anyone can make things, if they will focus their intent fine enough, but it is not everyone who can make things *create* themselves as I do. If you could see down into the depths as far as I can, you could see all sorts of creatures coming into being. Some of them go into this world, and some into other worlds and other times. *They* decide that as they take shape.

"Now you went out of your world," she said to Bertie, "and want to get back, which normally cannot happen, except that you have a purpose, which is like the Great Purpose of creative intent, which can bend space and time.

"Always keep the focus of your intent, and you may yet succeed."

"And if I do not?" said Bertie.

"Then you will remain in fairyland, never growing old, one of the eternally young."

"But we're here to grow up," said Tom manfully. "That's why we have come this far."

And then Mother Carey fell silent, and as they watched her blue eyes turned to ice, and her body became translucent, like an ice sculpture, and she seemed to have life for them no more.

Then they were no longer sure of the way that they had known when her eyes were upon them, and each thought it to be in a different direction, so that they were in danger of setting out north, east, south, and west. Finally they recognized that they must stick together, or be lost, and only one way permitted that—down!

And so they began diving again, this time straight down under Mother Carey's throne of creation. As they descended, they passed her creatures coming into being, some so strange and nightmarish that they knew that their destiny intended them for other worlds.

For seven days and seven nights they descended, but they no longer knew time, or up or down. When they came to the bottom, there was the great sea-serpent himself, coiled around the pillar that supported Mother Carey's throne. His head lay on the bottom with his mouth open to catch the unwary, but his mouth was so huge that it appeared to be a cave into which our unsuspecting friends wandered.

They took the serpent-dragon's teeth for icicles growing from the roof of the cave, and they thought his tongue to be a river of molten rock flowing out of the cave. When the mouth of this abyss suddenly crashed closed, they realized too late their mistake.

But fortune, or Mother Carey, was even then smiling upon them because the belly of the serpent was the beginning of Nowhere, and they now could proceed on to the End-of-Nowhere.

Beyond the cave-mouth of this fantastic creature, the way opened up into a strange countryside, full half a mile wide, not unlike what you might find in your own part of Kent or Cornwall except for the eerie glow which came from the roof-sky.

At first they walked through Waste-paperland, where all the stupid books lie in heaps, up hill and down dale, like leaves in a winter wood; and there they saw people digging and grubbing among them, to make worse books out of bad ones.

Then they went by the sea of slops, to the mountains of messes, and the territory of tuck, where the ground was very sticky, for it was made of bad toffee (not English toffee, of course).

Next they saw all the little people in the world, writing all the little books in the world, about all the other little people in the world, probably because they had no heroes to write about. But Tom and Bertie thought they would sooner have a jolly good fairy-tale about Jack the Giant-killer or Beauty and the Beast, which taught them something that they didn't know already.

Then they came to a very strange land, indeed, where everyone knew his neighbor's business better than his own. What a sight to see ploughs drawing horses, nails driving hammers, birds' nests taking boys, books making authors, bulls keeping china-shops, and blind bumblers installed as school principals.

Presently they heard voices behind calling "Stop!" Looking round they saw policemen's truncheons, or billy-clubs or night-sticks as you call them in America, running along without legs or arms. They were past being astonished now, by the wonders they had so far seen, so they stopped, and the chief club came up and asked their business. He had one eye in the middle of his upper end, so that when he looked at anything, being so stiff, he had to incline himself, till it was a wonder he did not tumble over, as he looked up and down, Tom, Bertie, Bongo and the little dog.

When they told him they were seeking Grimes at the Other-end-of-Nowhere, he replied that they were just about there.

And then he added, "I had better go along with you, and see you get to the right place. It's not far, though."

"Why have you no policemen to carry you?" asked Bertie.

"Because we are not like those clumsy-made truncheons in the other world which cannot get about without having a whole man to carry them. We do a policeman's work on our own, and do it very well, though I say it who should not."

"Then if you don't need a man to carry you," said Tom, "why have a leather thong on your handle."

"Why, to hang ourselves up, of course, when we are off duty."

The boys had their answer to that one, and a very good answer it was. They said no more, but jogged along at his side as he kept up a military pace. Ahead they saw the outline of a huge building, or groups of buildings, which stretched as far as the horizon and beyond.

Each building had a huge chimney that seemed to climb ninety miles high.

"Behold, the Other-end-of-Nowhere," said the truncheon, pausing.

Now the time, if there were such in this strange land, neared sunset, and near the horizon clouds of all colors and descriptions gathered. Then a curious thing happened.

The sun, which then should have plunged below the horizon, set like a chicken on its nest on the line of the horizon and moved not. But the clouds, which seemed to gather strength from its rays, fanned out in fantastic shapes towards the chimneys of each building. Slowly they took form—human form!—with elongate bodies and huge heads out of which glared the most doleful eyes human sight has ever seen. One after another, these smoke wraiths wound down the chimneys as if beckoned by a vortex of terrible energy.

For the first time in their journey, Tom and Bertie experienced a sense of evil, and with it came terror. The little dog and Bongo began to shiver uncontrollably.

Bertie turned to the truncheon and asked, "What place is this?"

"Why, these buildings all belong to Mr. Hitler."

Then a dark speck appeared across the blood-red half-eye of the sun. At first it seemed to be a bird, but then a noise of engines accompanied its flight. It had short stubby wings and flames emitted from one end. As it flew nearer, Bertie saw emblazoned on the wings the dread black and red swastika insignia. Then its motor stopped, and for a moment it glided on its own momentum.

Bertie stopped and picked up Bongo, and Tom did the same with the little dog. "Its time has come," said Bertie, "its clock has stopped ticking."

The flying bomb began its descent, dropping straight for one of the buildings. There was a terrible explosion, and then silence.

"You came just in time for the fireworks," said the truncheon. "Happens like that every day."

"What does it mean?" asked Bertie.

"Coming home to nest. Mr. Hitler made them, and one comes back to him at the end of each day.

"What about Grimes?" inquired Tom.

"Oh, he works up one of the chimneys. Mr. Hitler has to stoke the fires very hot to draw the smoke back down the chimney, and with it comes the soul that once went up the chimney.

"When the soul first goes up it leaves behind all the soot and evil acquired in life. Grimes and others like him clean the chimneys so that when the soul comes back to start a new life it has a fresh start."

They had been walking towards the great iron door of the nearest building. And there the club knocked twice, with its own head.

A wicket in the door opened, and out looked a tremendous old brass blunderbuss charged up to the muzzle with slugs, who acted as the door guard.

"What case is this?" he asked in a deep voice, out of his broad bell-mouth.

"If you please, sir, some young gentlemen come with Mother Carey's permission to visit Grimes, the master-sweep."

"Visitors. No one comes to visit *here*," said the blunderbuss. "Least not of their own accord."And here he laughed heartily. "Wait a min—I'll check the record."And he pulled in his muzzle, perhaps to look over the prison-lists. In a moment the great iron door swung open, and our friends were ushered in.

"Grimes is up chimney 345, so the young gentlemen had better go on to the roof and we'll have Grimes climb up to you."

"Very good, sir," said the truncheon to the blunderbuss.

"But it will be no use. Grimes is the most unremorseful, hard-hearted, foul-mouthed fellow I have working here; and he thinks of nothing but beer and pipes, which are not allowed here, of course."

At last they came to chimney number 345. Out of the top of it, his head and shoulders just showing, stuck poor Mr. Grimes, so sooty, and bleared, and ugly, that Tom could hardly bear to look at him. And in his mouth was a pipe, but it was not alight, though he was pulling at it with all his might.

"Attention, Mr. Grimes," said the billy-club, "here are some young gentlemen come to see you."

But Grimes said only bad words in response, and kept grumbling, "My pipe won't draw! My pipe won't draw!"

"Keep a civil tongue!" said the truncheon, and just like Punch hit Grimes such a crack over the head with itself, that his brains rattled inside like a dried walnut in its shell. He tried to get his hands out to rub the place where he hurt, but he could not, for they were stuck fast in the chimney.

119

"Hey!" he said. "Why, it's Tom! I suppose you have come here to laugh at my plight."

"That's not so," said Tom. "My friends and I have come a very long way to help you."

"Well, I don't want none of your help, unless you can find me a beer, or a light to this bothering pipe."

"I'll get you one," said Tom, and with two sticks he picked up a live coal, for there were plenty lying about, and put it in Grimes' pipe.

"It's no use," said the truncheon, leaning up against the chimney. "His heart is so cold that it freezes everything that comes near him."

"But can't I help you get out of the chimney?" asked Tom.

"No," replied the truncheon. "He has come to the place where he must help himself."

"Oh, yes," said Grimes sarcastically, "of course, it's *my* fault. Did I ask to be brought here to sweep Mr. Hitler's foul chimneys? Did I ask to have lighted straw put under me to make me go up? Did I ask to get stuck fast here in this chimney, because it was so shamefully clogged with soot? Did I ask to stay here for what seems like an hundred years and never get my pipe, nor my beer."

"No," answered a solemn voice from behind them. "Nor did Tom when you behaved the same way to him."

It was Mrs. Bedonebyasyoudid. When the truncheon saw her, it came to attention, then made a low bow. Tom and the others bowed too.

"Oh," said Tom to her, "please don't punish him anymore for what he did to me. The past is over and done with. Mayn't I try to get some of these bricks away, that he may move his arms?"

"You may try, of course," she said.

So Tom pulled and tugged at the bricks, and Bertie and Bongo tried to help him too, but the bricks would not budge. Then Tom took out the clean white handkerchief he had been saving for when he caught cold, and with it he tried to wipe the soot from Grimes eyes and face. And as he did so, he couldn't help smiling when he thought of all the times Grimes had sent him up the flues with kicks and blows, and how his eyes were blinded with soot. Now here he was wiping Grimes' eyes, and he smiled again.

"Making fun of me are ye?" said Grimes. "If I had an arm free now, I'd learn ye!"

Mrs. B. just shook her head in irritation with Grimes' reaction to Tom's kind act. Despite Tom's effort, none of the grime would come off, and his handkerchief remained as clean as when he started.

"Oh, dear," he said. "I have come all this way, through all these strange places, to help you, and now I am no use after all."

"You had best leave me alone now," said Grimes. "You are a kind-hearted, forgiving little chap, and that's the truth, but you'd best be off. If I could make it up to you, I would. If I was to do it again, how different I would be. I'd treat you like my own son. And you coming all this way to help me like I was your own father. But I've made my bed, and I must lie in it. Foul I would be, and foul I am, as an Irishwoman once said to me. It's me own fault, and it's too late for anyone to help, so you'd best be off."

And he hung his head and began blubbering like a great baby, till his pipe dropped out of his mouth and broke all to bits.

"Never too late," said the fairy, in such a strange soft new voice that Tom looked up to see that she had become her beautiful sister, Mrs. Doasyouwouldbedoneby.

Nor was it too late. For, as poor Grimes blubbered on, his tears of repentance did for him what nobody else on earth could do; for they washed the soot out of his eyes and off his face and clothes, and then they dissolved the mortar that held the bricks in place, and Grimes was free of his self-made prison.

"Well," said Mrs. Doasyouwouldbedoneby, "you finally got that chimney clean!"

So Grimes stepped out of the chimney, and really, if it had not been for the scars on his face and hands, he looked as clean and respectable as a master-sweep need look.

"Take him away," said the fairy to the truncheon, "and give him his ticket to Mother Carey's throne."

"What's to become of me now?" asked Grimes somewhat uneasily.

"Why, you start all over—born again," said the fairy.

"It's not bad," said Tom. "We saw everything coming into creation as we swam down here."

Bertie raised his hand.

"Yes, my dear."

"Please, mam," he said. "May I tell Mr. Grimes Mother Carey's secret?"

"Would you tell a great secret?" she said, and her eyes changed to the dazzling azure color of Mother Carey herself. Then Bertie knew that in that gaze he had been given permission.

"The secret is this," he said to Grimes. "You make yourself what you want to be."

At this Grimes did a little jig of joy.

"Oh, to be a little chap again in Vendale where I came from! To play in the clear brook under the appleboughs, and to taste that fruit again! And Tom, do you remember the limestone spring where I washed my head on our way to Harthover Hall?! I must find it again! How clean I will be."

"How clean you will be, if you wish to be," said a new voice, and, behold! the Irishwoman of the spring stood before them.

"I gave you your warning then," she said. "Every cruel and mean thing that you did—you were disobeying me."

"If only I'd known, mam—"

"You knew well enough. But your work is done here, and Tom's too. The chimney is clean."

Then there were leave-takings, for Grimes wanted to know if Tom could be created anew with him, so that they could be chums for life, but the fairy told him, "Tom has not yet grown up."

So Grimes parted from the company, under the watchful eye of the truncheon, with a last remark to Tom.

"This time I don't think I'll make myself a sweep," he said. "No, the King of England is a good job for me," and he laughed heartily. "Come and see me at the palace, Tom," and he was gone.

The Irishwoman went to the chimney and looked down. "Such a nice clean chimney," she said. "Many souls will be using it on the way to Mother Carey's throne."

"The souls come in that went out that chimney?" asked Bertie.

"Yes," she said.

"What happens to the soul when it leaves and when it comes back?" asked Tom.

"Well, when it left *this* chimney, it deposited all its blackness, all the grime, soot, ignorance, and evil, on the walls of the chimney. Then it goes to a place, which really isn't a place but a state of being, where it receives instruction in the Light. It learns of its potentials as a part of the Light. Then, when it is ready to begin another life-lesson, it returns. Those souls that went up these chimneys return here when the chimney is clean, because they went *up* the chimneys because of Mr. Hitler's evil. He must clean them—with the help of many, many men like Grimes—because a relation to the soul was created when he killed that soul. So he must make the chimney clean for the return passage to Mother Carey's throne."

"Then time runs backwards," said Bertie.

"Oh, it runs backwards, to be sure," said the fairy, "and inside-out, and perpendicular, and many, many more ways than you can imagine. Remember, time doesn't have hands like a clock, which only *measures* time on earth, and wouldn't work in other universes where time doesn't always move forward.

"Remember, too, there is no time to a soul. They have as much time as they wish to *unlearn* the last life-lesson, and to prepare for the next life-lesson. Remember, always, Light is everywhere at *once*. *Once* has no past or future, no time or place.

"Now your work is done, Tom, but Bertie's is just beginning, so I shall take you four up my backstairs, the shortcut to the world. Cover your eyes and don't peek."

And in a twinkling of an eye they were back and Ellie stood before Tom.

"Oh, Miss Ellie," she said, "how you are grown!"

"Oh, Tom," said she, "how you are grown, too!"

And no wonder; they were both quite grown-up—he into a tall man, and she into a beautiful woman.

"You may take him home with you now on Sundays, Ellie," said the fairy. "He has won his spurs in the great battle, and become fit to go with you and be a man, because he has done the thing he did not like."

"But who is the little boy, with the monkey and the dog?" asked Ellie.

"Why, I remember them, I think," said Tom uncertainly.

"But why are they naked," said Ellie, "and why are they wearing those frilly lace collars?"

Now Bertie and Bongo blushed because they were still water-babies.

122

123

"I don't want to be a baby anymore," said Bertie.

"Wait," said Tom, "now I remember. These were my friends on my big adventure many years ago. Ellie and I will be your father and mother, and you can grow up with us."

"I would like that," said Bertie. "How about you, Bongo?"

"Aren't you forgetting something, my little dear?" said the fairy, and she gazed deeply into his eyes. Once again he saw the clear azure depths of Peacepool and the all-knowing eyes of Mother Carey, and then he remembered his Purpose.

"I want to grow up and be somebody like Tom," said Bertie.

"Then you shall have to go out of this story and into the next," said the fairy, "for you don't belong in it, and, therefore, can not grow up here. But I shall help you find you way into the next story."

Then Bertie picked up Bongo and went to Tom to shake hands and say goodbye, but his lip began to quiver, and it was all he could do to keep from crying. So they said goodbye, and Tom went home with Ellie on Sundays, and sometimes on weekdays too; and he is now a great man of science, and no one knows except you and I that the cradle for the ideas that were to make him famous came when he was a wee water-baby under the sea.

"And what became of Tom's dog?"

"Oh, you may see him any clear night in July trying to drink from the Big Dipper; for the old dog-star was so worn out by the last three hot summers that there have been no dog-days since, so they had to take him down and put Tom's dog up in his place. Therefore, as new brooms sweep clean, we may hope for some warm weather this year. And that is the end of my story.

"Or, at least, that was where my story ended before, but now I am stuck with two characters who weren't in my story before, and I don't know how to get rid of them."

"I'll help you," said the fairy to the author.

"How can a character in a book talk to the author?" asked Bertie.

"All things are possible in fairyland," said the fairy.

"Then I'll leave it to you," said the author. "Besides, I have to prepare next Sunday's sermon—June 12, 1874—for my parish at Eversley in Hampshire, so I really do not have any more time to devote to *The Water-Babies*."

"Remember that time is relative," said Bertie. "Well, we'll be getting along with the fairy. It has been a pleasure meeting you and being in your story. Don't worry, we won't be *permanently* there in the future, only just for this one time."

"I don't mind," said the author, graciously. "It was a pleasure having you. I sincerely hope everything turns out all right for you, and that you accomplish your purpose."

"Thank you," said Bertie. "I think we'd better be going now—I see the fairy is moving off the page.

"Oh, Mr. Kingsley—one last thing," said Bertie. "Know you that your story becomes a classic. I read it in 1950."

"Nineteen fifty!" said Kingsley. "Well, I shall give thanks to God." And he turned his hand to his sermon.

The fairy beckoned, and Bertie and Bongo followed. They went back into the sea and swam along with her until she came to the end of the sea of *The Water-Babies*.

"This is as far as I can go," she said, "for I don't belong in the next story."

Bertie could see no seam between the water they were in and that which lay ahead. But now in the distance, an island began to manifest, and a ship at anchor. From its masthead flew the unmistakable skull and crossbones of the Jolly Roger.

"Look, Bongo," said Bertie, "pirates!"

"The time has come for you to begin growing up," said the fairy. "Fairy love will be with you, where and when you go, and fairy eyes will be upon you. Now swim straight for the island, and never look back. You won't have gills anymore, so keep your heads above water, lads. Tom you may see—and his little dog too—guiding you whenever you look above."

"I'll miss you," said Bertie, and he began to feel a sad quiver. But he knew the time had come to begin growing up, so he struck out bravely towards the distant island.

As they swam they heard the good fairy's voice weaving a kind of fairy spell or song, that the waves sang back to them. Long, long, they swam, the song ever with them. Ahead, a sand bar curved round them to right and left, and in the middle a narrow estuary, into which they swam out of the ocean and into the next story, their gills gone and their lungs gulping air.

GLOSSARY
BOOK TWO

abyss: a deep space

amphibious: able to live on both land or water

antennae: movable feelers

blunderbuss: an old-fashioned gun

body-surfing: riding a wave without a surfboard

bogy-painters: artists who paint frightening pictures

brachiapod: a selfish like a clam

calomel and jalap, etc: powders to move the bowels

collided: came together with violent impact

colloquialism: an expression not of formal or proper speech

compassionate: full of caring

conjuring tricks: producing effects as if by magic

coracle: a small boat covered with skin or cloth

doleful: full of woe or sorrow

dosed: gave a portion of medicine

emblazoned: painted in color

emetic: a medicine to make one throw up

enlightenment: a state of being given spiritual light or knowledge

estuary: an inlet or narrow body of water

fastidious: particularly careful

foam-bedizened crests: wave tops adorned with a showy or spectacular display of foam

gobbling: eating very fast

grammatical: proper use of the parts of speech

grenadier: a soldier

grotto: a cave

Gulliver: the hero of a book by Swift. Gulliver visited imaginary countries of giants and tiny people

hullabaloo: an uproar or noisy disturbance

insignia: a distinguishing mark or sign

intertwine: interwoven or linked together

Jolly Roger: the flag of pirate ships bearing a white skull and bones against a black background

lochs and friths: narrow arms of the sea

manorial: of a manor or large estate of land

mirth: laughter

moralizing: making distinctions between right and wrong

nautical: of the sea or seamen

nymphs and muses: goddesses of mythology

obstinacy: stubbornness

perils: dangers or risks

ploughs: or plows. a tool used for cutting lifting, and turning the ground

poaching: to trespass in order to hunt or steal animals

political economy: an organized system of government

protruded: stuck out

rueful: full of sorrow

sarcastically: spoken with a sneer or with irony

sea-urchins: a sea creature with a prickly shell

sedges: hollow-stemmed plants found in wet places

slag: cinder

snickered: laughed in a disrespectful way

spatial relations: relating to space

thrashed: to beat or whip

tranquil: peaceful

translucent: allowing light to shine through

trials and tribulations: trying or testing events and extreme troubles

trudging: to walk with great effort or wearily

twiddled: to twirl with the fingers

undulating: waving

vexation: the state of being irritated or annoyed

vortex: a whirling mass of water or air

BOOK THREE

AIR

Straight on till morning.

Feeling that Peter was on his way back, the Neverland had again woke to life. I ought to have used the pluperfect and say wakened, but I don't know why I should bother. Children these days would not know the difference.

Not like 1911, when one was properly schooled in grammar, and did not abuse routinely the King's English, or the Queen's either for that matter. Boys are worst of all, of course, harder on grammar than knees or a pair of sneakers, never able to resist the opportunity to fracture a phrase or double a negative.

I was not like that, of course, when I was a boy. In a drawer of my bureau, where I should have had clean socks, I kept a drawer full of semi-colons. I was forever taking them apart—much to my mother's dismay—for I could never get them back together again. I would use a dot for a period at the end of a sentence, and then be left with only a comma, which I saved for clauses, although Santa never required any.

And so, you see, the youthful education of a writer is somewhat different from that of other children, who cannot resist taking apart cars and clocks—anything mechanical, for that matter—but will resist to the death diagramming a sentence into its component elements.

Well, as I was saying, I ought to have used the pluperfect and say Neverland wakened, but woke is better and was always used by Peter.

In his absence, things are usually quiet on the island. The fairies dally an hour longer, painting the flowers with morning freshness; the beasts loll about heavy-lidded on the grasses; the redskins suspend the hunt for the pirates, content to sit around campfires sharpening hatchets and fixing heads to arrows; and when pirates and lost boys meet, they merely bite their thumbs in contempt at one another; for without Peter Pan, the prize whom Captain Hook sought, they were merely boys to the buccaneers, unworthy of manly combat.

On this evening the chief forces of the island were disposed as follows. The lost boys were out looking for Peter, the pirates were out looking for the lost boys, the redskins were out looking for the pirates, and the beasts were out looking for the redskins. They were going round and round the island, but they did not meet because all were going at the same rate of speed like figures on a merry-go-round.

All wanted blood except the boys, who liked it as a rule, but tonight were out to greet their captain, Peter. The boys on the island vary, of course, in numbers, according as they are killed, and so on, and when they seem to be growing up, which is against the rules, Peter thins them out; but at this time there were six of them, counting the twins as two.

Let us pretend to lie here in the concealing grasses and watch the lost boys pass, each with his hand upon his crotch, for they were forbidden to tinkle except at

Peter's command—when he rang the tinkle-bell, and he had been gone a very long time.

First comes Tootles, not the least brave, not the most brave, but middling brave. Adventures, however, had a nasty way of eluding him. Just when he had stepped round the corner to the sweetshop (no, that's not right; there aren't any shops on this island, sweet or nasty), an adventure would turn up right where he had been standing not a moment before. Of course, by that time it was too late, and then when he returned the other boys would be mopping up positively buckets of pirate blood. So where is the fun in that?How can one prove his mettle when adventure itself is afraid of one?

Next comes Nibs, the debonair, followed by Slightly, who cuts whistles out of the trees and dances ecstatically to his own tunes. Slightly is the most conceited of the boys. He thinks he remembers the days before he was lost, with their manners and customs, and this has given his nose an offensive tilt. Curly is fourth; he is a pickle, and so often has he had to deliver up his person when Peter said sternly, "Stand forth the one who did this thing," that now at the command he stands forth automatically whether he has done the thing or not. Last come the Twins, who cannot be described because we should be sure to be describing the wrong one. Peter never quite knew what twins were, and his band were not allowed to know anything he did not know, so these two were always vague about themselves, and did their best to give satisfaction by keeping close together in an apologetic sort of way.

The boys vanish in the gloom, and after a pause, but not a long pause, for things go briskly on the island, come the pirates on their track. We hear them before they are seen, and it is always the same dreadful song:

"Avast belay, yo ho, heave to,

A-pirating we go,

And if we're parted by a shot

We're sure to meet below!"

A more villainous-looking lot never hung in a row on Execution dock. Here, a little in advance, ever and again with his head to the ground listening, his great arms bare, pieces of eight in his ears as ornaments, is the handsome Italian Cecco, who cut his name in letters of blood on the back of the governor of the prison at Goa. That gigantic black behind him has had many names since he dropped the one with which dusky mothers still terrify their children on the bank of the Guidjo-no. Here is Bill Jukes, every inch of him tattooed, the same Bill Jukes who got six dozen on the *Walrus* from Flint before he would drop the bag of gold coins, and Cookson, said to be Black Murphy's brother (but this was never proved); and Gentleman Starkey, once an usher in a public school and still dainty in his ways of killing, and Skylights (Morgan Skylights); and the Irish bo'sun Smee, an oddly genial man, who stabbed, so to speak, without offense, and was the only Nonconformist in Hook's crew; and Noodler, whose hands were fixed on backwards; and Robert Mullins and Alf Mason and many another ruffian long known and feared on the Spanish Main.

In the midst of them, the blackest and largest jewel in that dark setting, reclined James Hook, or, as he wrote himself, Jas. Hook, of whom it is said he was the only man that the Sea-Cook feared. He lay at his ease in a rough chariot drawn and propelled by his men, and instead of a right hand he had the iron hook with which ever and anon he encouraged them to increase their pace. As dogs this terrible man treated and addressed them, and as dogs they obeyed him. In person he was cadaverous and blackavised, and his hair was dressed in long curls, which at a little distance looked like black candles, and gave a singularly threatening expression to his handsome countenance. His eyes were of the blue of the forget-me-not, and of a profound melancholy, save when he was plunging his hook into you, at which time two red spots appeared in them and lit them up horribly. In manner, something of the grand lord still clung to him, so that he even ripped you up with an air, and I have been told that he was a storyteller of repute. He was never more sinister than when he was most polite, which is probably the truest test of breeding; and the elegance of his speech, even when he was swearing, no less than the distinction of his manner, showed him one of a different cast from his crew. A man of indomitable courage, it was said of him that the only thing he shied at was the sight of his own blood, which was thick and of an unusual color. In dress he somewhat aped the attire associated with the name of Charles II, having heard it said in some earlier period of his career that he bore a strange resemblance to the ill-fated Stuarts; and in his mouth he had a holder of his own contrivance, which enabled him to smoke two cigars at once. But undoubtedly the grimmest part of him was his iron claw.

Let us now kill a pirate, to show Hook's method. Skylights will do. As they pass, Skylights lurches clumsily against him, ruffling his lace collar; the hook shoots forth, there is a tearing sound and one screech, then the body is kicked aside, and the pirates pass on. He has not even taken the cigars from his mouth. Good form, Hook!

Such is the terrible man against whom Peter Pan is pitted. Which will win?

On the trail of the pirates, stealing noiselessly down the war-path, which is not visible to inexperienced eyes, come the redskins, every one of them with his eyes peeled. They carry tomahawks and knives, and their naked bodies gleam with paint and oil. Strung around them are scalps, of boys as well as pirates, for these are the Piccaninny tribe, and not to be confused with the softer-hearted Delawares, or the Hurons. In the van, on all fours, is Great Big Little Panther, a brave of so many scalps that in his present position they somewhat impede his progress. Bringing up the rear, the place of greatest danger, comes Tiger Lily, proudly erect, a princess in her own right. She is the most beautiful of dusky Dianas and the belle of the Piccaninnies, flirtatious, cold, and amorous by turns; there is not a brave who would not have the wayward thing to wife, but she staves off the altar with a hatchet. Observe how they pass over fallen twigs without making the slightest noise. The only sound to be heard is their somewhat heavy breathing. The fact is that they are all a little fat just now after the heavy gorging, but in time they will work this off. For the moment, however, it constitutes their chief danger.

The redskins disappear as they have come, and soon their place is taken by the beasts, a great and motley procession: lions, tigers, bears, and the innumerable smaller savage things that flee from them, for every kind of beast, and, more particularly, all the man-eaters live cheek by jowl on the favored island. Their tongues are hanging out, they are hungry tonight.

When they have passed, comes the last figure of all, a gigantic crocodile. We shall see for whom she is looking presently.

The crocodile passes, but soon the boys appear again, for the procession must continue indefinitely until one of the parties stops or changes its pace. Then quickly they will be on top of each other.

All are keeping a sharp lookout in front, but none suspects that the danger may be creeping up from behind. This shows how real the island was.

The first ones to fall out of the moving circle were the boys. They flung themselves down on the grass, close to their underground home.

"I do wish Peter would come back," every one of them said nervously, though in height and still more in breadth they were all larger than their captain.

Ah, but when Peter came flying in, from London or wherever his flight plan took him, the lost boys became a force to be reckoned with, and if you put your ear to the ground now, you might get an earful of sand, or hear the pit-pat of many feet, bare or moccasined, human or hooved, as the island readied for frenzied combat.

"I am the only one who is not afraid of the pirates," Slightly said in a tone that prevented his being a general favorite; but perhaps some distant sound disturbed him, for he added hastily, "but I wish Peter would come back, and tell us whether he has heard anything more about Cinderella."

They talked of Cinderella, and Tootles was confident that his mother must have been very like her.

It was only in Peter's absence that they could speak of mothers, the subject being forbidden by him as silly.

"All I remember about my mother," said Nibs, "is that she often said to father, 'Oh, how I wish I had a checkbook of my own.' I don't know what a check-book is, but I should love to give my mother one."

We take leave now of this childish prattle to move to a more sinister scene—the deck of the pirate ship *Jolly Roger*. Bill Jakes is on watch in the crows nest. A distant splashing has preoccupied his gaze, and now he calls to Smee below to bring Captain Hook on deck. He has been playing the harpsichord, as is his wont on calm evenings when the ticking of a clock inside a crocodile is easily discerned, and his mood at being disturbed is as black as his dreadlocks.

"Someone's nose wants wiping with my hook?" he asked no one in particular.

"No, captain," Smee replies apologetically. "The watch has spotted something swimming this way, and it looks to be one of those boys you hate."

"Where?" He turns in the direction indicated and pulls a spyglass from his pocket. He sights steadily for a moment, and then collapses the glass back into his pocket.

"A boy, all right, and something smaller alongside, but they're coming from the direction of the open sea."

It was at this moment that Bertie and Bongo stopped swimming to rest. Treading water, for their gills had left them the moment they had swum out of the last story's sea, they scrutinized every detail of the ship. The limply flying flag of the skull and crossbones had a foreboding look, as did the huge cannon, affectionately called Long Tom by the pirates.

"Do you think we should swim to the ship?" Bertie queried Bongo.

Bongo shook his head slowly and very emphatically from side to side, in a gesture that meant "no way."

Bertie agreed. "Then it's on to the island." And they began swimming away from the ship.

Satisfied that there was no danger, Hook brought out his dual-cigar holder and lit up, and leaned his back against the mast. Smee sidled up to him.

"We could part their hair with a shot from Long Tom, if ye likes, Captain."

"Swatting flies with iron hammers! They are only two, and I want to mischief all seven."

Hook heaved a heavy sigh; and I know not why, perhaps it was because of the soft beauty of the evening, but there came over him a desire to confide to his faithful bo'sun the story of his life. Smee had endeared himself to the crew by working (at a sewing machine) when his other duties were through, in order to repair the crew's clothes. Now he sat down at it, and the reassuring whir began.

As usual, Hook was in the depths of a depression. Calm nights seemed to bring them on, but talking to Smee always helped scatter his personal gloom. But this inscrutable man never felt more alone than when surrounded by his sea-dogs. They were socially so inferior to him.

Hook was not his true name. To reveal who he really was (even at this date) would set England into a turmoil; but as those who read between the lines must already know, he had been at a famous public school; and its traditions still clung to him like garments. He yet adhered in his walk to the school's distinguished slouch. But above all he retained the passion for good form.

Hook knew that it was bad form to be constantly thinking about good form. But he knew too that in living the life of a pirate, he played the blackest of all roles. Were he to be victorious over all his adversaries, and the British Navy to boot, it would be the poorest of all form. History only immortalized the triumph of Good over Evil; all else was swept under the historical carpet.

Therefore, the double bind had him fast. He could not win. He had to lose eventually. Anything else would be damning bad form. But if he could not beat the system, he

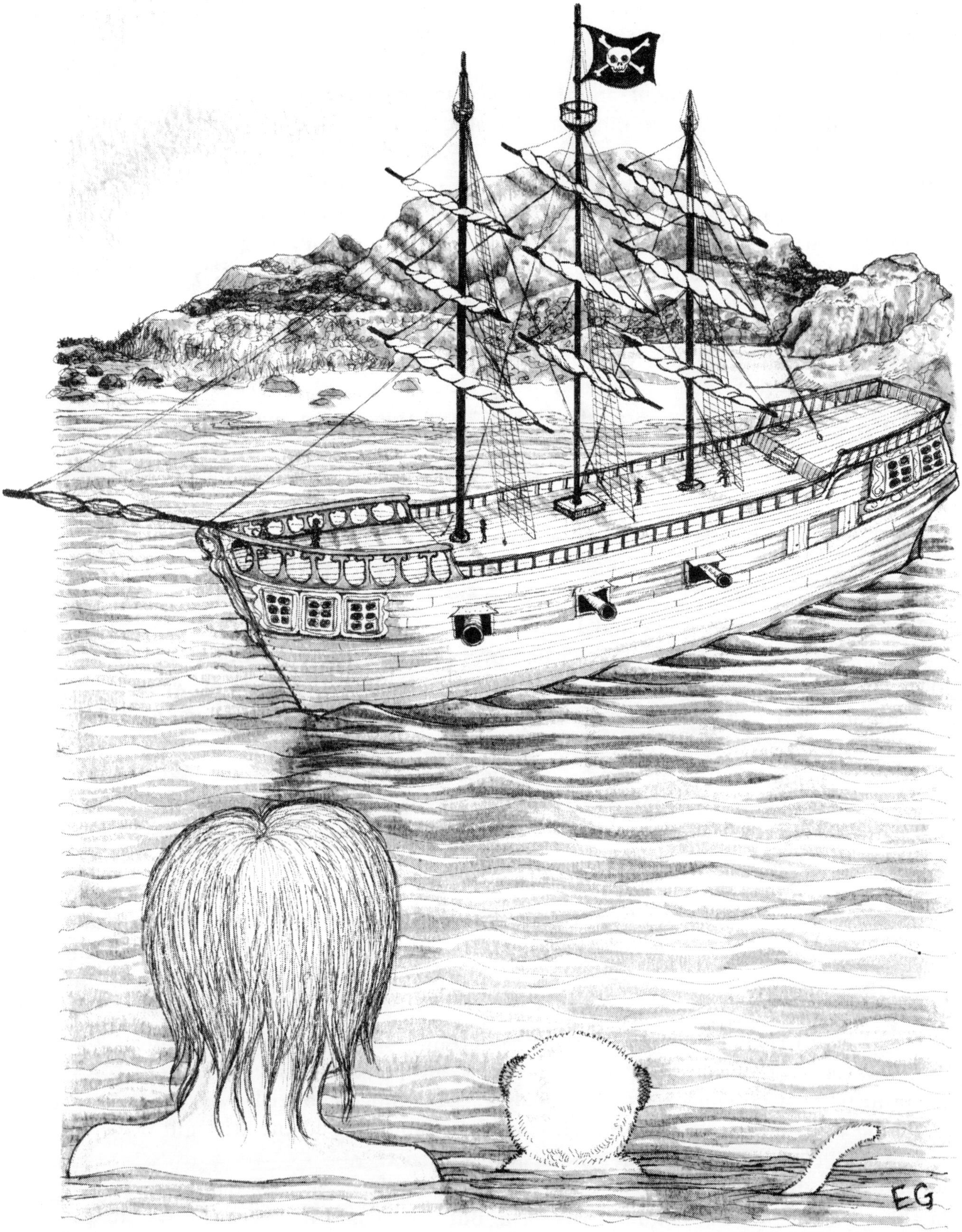
EG

wanted above all else to destroy Peter Pan. Then he would take his medicine however bitter it might be. It was the boy's unbearable cockiness that infuriated him. Pan was clever, no doubt, to have eluded him so far, but did he have to crow about it. That was the worst of bad form. To do well without humility nullified the doing.

As he worked at his mending, Smee only half-listened to his Captain. Anon he caught the word Peter.

"Most of all," Hook was saying passionately, "I want their captain, Peter Pan. 'Twas he cut off my arm." He brandished the hook threateningly. "I've waited long to shake his hand with this. Oh, I'll tear him. Pan flung my arm," he said wincing, "to a crocodile!"

"I have often noticed," said Smee, "your strange dread of crocodiles."

"Not of crocodiles," said Hook angrily, "but of that *one* crocodile." He lowered his voice, "It liked my arm so much, Smee, that it has followed me ever since, from sea to sea and from land to land, licking its lips for the rest of me."

"In a way," said Smee brightly, "it's sort of a compliment."

"I want no such compliments." Hook barked petulantly, "I want Peter Pan, who first gave the brute its taste for me."

He placed an arm over Smee's shoulder and drew closer, nearly whispering in his ear. "Smee, that crocodile would have had me before this, but by a lucky chance it swallowed a clock, which goes tick-tock inside it, and so before it can get to me, I hear the tick and run away."

"Some day," said Smee, "the clock will run down, and then the crocodile will have his Hook."

Hook paused and wet his lips. "Curious that you should say that Smee. In the Egyptian mythology, at the end of time there is a judgment of the soul, a weighing against a feather. Those souls who do not pass the judgment and are found unworthy are fed—can you guess Smee?"

"To a crocodile."

"Aye, aye," said Hook.

When Bertie and Bongo veered away from the pirate ship, their approach took them to Mermaid's Lagoon, certainly the loveliest spot in all of Neverland. Night had come on, and the full moon was just rising when Bertie and Bongo paused to rest on Marooner's Rock at the mouth of the lagoon. Marooner's Rock was so called because evil captains had chained sailors there, leaving them to drown when the tide rose submerging the rock.

As our friends rested, determined that night to finish the last quarter mile of their swim to the shore, there came a curious slapping of fins in the water, as if a school of large fish were swimming by. Then the water around the rock began to boil, and Bongo climbed into Bertie's lap for safety.

Presently the head of a beautiful woman appeared, and then another. They sat on the ledges of the rock, and began—oh, wondrous to behold—to comb their long green hair. And as they combed they sang songs in a language Bertie had never heard, nor would he hear ever again. All this time, they stared at Bongo. Perplexed, he kept looking around at Bertie, as if to ask why.

"I think they like you," said Bertie. This made Bongo embarrassed, and he blushed and looked down at the rock.

Soon the mermaids were not content just to look at Bongo; they began to crawl out of the water onto the rock to have a closer look at him. Then Bertie saw that from the waist down they had fish-tails, with glistening scales of blue, green, and aquamarine.

"They are mermaids," said Bertie. Bongo had folded his hands in his lap, and now nodded his head up and down, as if to say because they were mermaids, their overtures to him were to be excused.

Perhaps because they had never seen a monkey before—the island had none—or perhaps because he resembled a mermaid fetus, Bongo had them all enchanted.

Now their luring song became more insistent, high-pitched and shriller, and it was obvious that they wanted Bongo to join them in the water. Bertie held him tighter, and they began to regard him with malice in their eyes.

Bongo was not afraid for himself, but he worried that they might harm Bertie if he held on to him, so he turned around and gave him a look that said, "Perhaps I'd better." Bongo, as you know, became an expert surfer in our last story, and had been watching the waves beginning to break on a submerged ledge of rock. As the tide receded further, this break became greater.

Just then Bongo noticed a large set of waves moving from sea towards the rock. Standing up he gave a kind of "Hi girls" wave with his paw. A thrilling sigh shivered through the whole school of mermaids. As the biggest wave passed over the ledge, Bongo jumped off the rock with all the mermaids in pursuit. He caught the wave high up with perfect form and rode it towards the center of the lagoon. Some of the lucky mermaids saw what he was doing and plunged into the same wave, vigorously flapping their tails. To their surprise and joy, they were lifted up and carried along too, until the wave spent itself in the center of the lagoon.

Then you never heard such shrieking and laughter as came from the ecstatic school. Bertie sat on the rock, admittedly jealous, as Bongo and the beauties went back and forth across the lagoon on the moonlit surf.

Finally, somewhat fed-up, Bertie stood up and called to Bongo.

"It's time we were getting into shore!"

Bongo nodded obediently, and as if with one voice, the school emitted a mournful cry that sounded like "Ohhhhhhh!", turned-turtle in the water, and were gone. Nary a ripple betrayed their passing.

As Bertie and Bongo swam silently towards shore, was it a smile that began on Bongo's face, or was it Bertie's imagination? He was not sure, nor can we be.

At last the shore was reached, but not without danger, for the bushes at the very edge of the lagoon concealed shifting shapes that melted into the moonlit landscape and then reappeared. Finally a figure stepped from the brush and confronted them.

"Who goes there?!"

"Bertie and Bongo."

"Who?"

"Bertie and Bongo."

Then snickering and stifled laughter, which could not be contained. The Lost Boys gathered around them, fingering their knives as if about to use them.

"Are they twins?" said the Twins.

This started the boys giggling again.

Finally Slightly spoke up. "Come on, chaps, we have to sort this out before Peter gets back. He'll want to know who they are."

"Peter who?" said Bertie.

"Peter Pan," said the boys as one.

"Oh, that explains the pirate ship," said Bertie to Bongo. "We're in the *Peter Pan* story."

"What story?" said Nibs. "This is serious."

"Then you don't know you're in a story," said Bertie. "The animals did."

"What animals?" said Curly.

"The animals in *The Wind in the Willows*."

"Now look here," said Slightly, "you've landed on our island, and we want to know who you are and why you are here?"

"I don't know why we're here," said Bertie, "but we come from London, 1950."

"Nineteen *fifty*!" said the Twins.

"What were you doing out on Marooner's Rock?" Tootles asked.

"Just meeting some of the native girls," said Bertie casually.

"We saw what you were doing," Slightly said. "You were swimming with them."

Bongo shook his head emphatically.

"No, my friend says he was *surfing* with them," corrected Bertie.

"What's surfing?" they all asked at once.

"We'll show you tomorrow," replied Bertie, "but we really need to get some sleep now. Can we stay with you?"

"Well, they're obviously not Redskins," Nibs said.

"Nobody sleeps until Peter gets back," answered Slightly.

"Can he really fly?" asked Bertie.

"Can mermaids swim?" replied Slightly contemptuously.

"Can bees sting?" added Curly.

"Can ducks quack?"

"Can Hook swear?"

"Can snakes crawl?"

"Can Tink pinch?"

135

"Can Bongo surf?" concluded Bertie.

"Stow it!" said Slightly commandingly. Now he took out his knife and began to finger it, running his thumb up and down the tip. Advancing on Bertie, he said, "Look, if you won't talk I can make you. Where did you come from just now?"

"From out to sea," said Bertie. "We swam in."

"What were you doing out there?" Slightly persisted.

"We were in *The Water-Babies*," Bertie answered honestly.

"What kind of a place is *that*?" Nibs asked sarcastically.

"It's not a place, it's a book," Bertie replied.

"But you were *in* it," Curly reasoned.

"Yes, we were," said Bertie. "Well, I guess it's a place too. Look, here *you* are in Neverland, and you're also in a book."

"What book?" said Slightly.

"*Peter Pan*," responded Bertie.

"Well, when Peter gets back, he will deal with you for all your smart answers, and he doesn't have just a knife."

"That's right," spoke Tootles, "Peter has a great big sword."

"'Hook or me this time!'" yelled Nibs, echoing Peter's battlecry.

"How come the rest of you don't have swords?" questioned Bertie. "I mean, after all, if you're going to fight the pirates shouldn't you be equal?"

"It's not allowed!" cried the Twins. "Only Peter can have a sword!"

"Who said?" questioned Bertie.

"Peter!" they all said as one.

"Look, if you're going to stay," said Slightly, "you might as well learn the rules."

"First," spoke up Nibs, "nobody touches Peter."

"Why not?" demanded Bertie.

"Because Peter makes up the rules, and that's one of them," Slightly replied angrily.

"And we always salute Peter," said Curly.

"All right," said Bertie. "If Peter is your leader, I can go along with that."

"And the other thing," added Tootles, "we can't talk about our mothers."

"No, Peter doesn't permit that," confirmed the Twins.

"Well," responded Bertie, "I'll talk about my mother if I want to."

"No, you won't," retorted Slightly. "Peter will stop your mouth."

Bongo shook his head emphatically in his "no way" gesture.

"We'll see about that!?" rejoined Bertie.

"All right, now shut up everybody," ordered Slightly. "Keep your eyes peeled for Peter."

Just then a terrific explosion rent the air.

"They fired Long Tom," said Nibs in awe.

Bertie felt some fingers fumbling for his hand, and heard Tootles whisper in his ear. "Could I please hold Bongo's other hand?" he said.

For what seemed an hour, they stood thusly, Bertie and Tootles, with Bongo in the middle holding their hands. Slightly had ordered the Lost Boys to spread out, and the others had vanished into the dense undergrowth. Then Bongo began to nod off, and Bertie had the bright idea that they climb a tree for sleep and safety.

Secure in its branches, Tootles said, "I have to keep watching for Peter, but since you're not really one of us yet, I don't suppose it matters if you and Bongo want to go to sleep."

So Bongo curled up on Bertie's chest with his head tucked under Bertie's chin, and soon they were fast asleep. They were happily dreaming of swimming with mermaids in the lagoon when they were awakened to hear Nibs calling from the ground below.

"I have seen a wonderfuller thing, Tootles. A great white Wendy bird. It is flying this way."

Even then they heard a kind of moaning cry, in which the words, "Poor Wendy" were repeated over and over again. Suddenly, right over the top of their tree, flew a girl in a white nightgown, pursued by a tiny, sparkling light that darted in and out against her exposed arms and legs.

"Why it's Tinker Bell," spoke Tootles, "and she's after the Wendy bird."

Seeing Tootles, Tinker Bell zoomed down in front of him, and Bertie heard words which were not so much speech but the melodious chiming of bells.

"Tootles," the bells said, "Peter wants you to shoot the Wendy."

Without hesitating, Tootles drew his bow and aimed. Adventure had at last come to him.

"That's not a bird! That's a girl!" said Bertie, spoiling Tootles' shot.

The arrow passed over Wendy's shoulder and struck Tinker Bell as she hovered about to pinch Wendy again.

"Oh, dear," said Wendy, "what has she done now?" as Tinker Bell hurtled to earth.

"You hit my arm!" accused Tootles.

"You can't shoot girls!" rejoined Bertie.

Wendy glided down to the spot where Tinker Bell had fallen, soon followed by four boys, one of whom was the leader of the Lost Boys, Peter Pan, who crowed triumphantly to announce his arrival. Tootles, Bertie, and Bongo clambered down from the tree. Now all the Lost Boys were standing in a circle around Wendy, Peter, and Tinker Bell, who did not move, her fairy light pulsing feebly.

"Who has done this?" Peter inquired of the surrounding circle. Curly automatically started to step forward, and then realized that this was one time he was not to blame. Timidly, Tootles spoke up, "I did."

"Why?" questioned Peter.

"She told me you wanted me to shoot the Wendy."

"Shoot Wendy?!" said Peter angrily, "I have brought her to be your mother."

"He didn't do it," spoke up Bertie. "I hit his arm."

"Who are you?" asked Peter.

"We have been trying to find that out ever since he arrived here," stated Slightly importantly.

"Well, then, what have you to say for yourself?" demanded Peter.

"My name is Bertie, and where I come from we don't shoot girls."

"Where *do* you come from?"

"Hampstead—London."

"I've just come from there—Kensington, actually."

"Yes," said Bertie, "there's a statue of you in the park."

"Where?"

"By the Long Water."

"I ran away to Kensington Park," said Peter. "I know every bush and every blade of grass. There is no statue of me, although it would be nice to think that there is."

"Liar!" yelled Slightly at Bertie.

"Don't ever call me a liar!" responded Bertie.

"All right, that's enough!" said Peter. "The question is, why did Tink tell Tootles to shoot Wendy?"

"I don't know," said Wendy, "but I did hear her say that to this boy, and the other boy—Bertie—hit his arm, and probably saved my life. Besides, look at my arms, she was pinching me all the way."

Wendy held out her arms for Peter to see. Peter knelt down by Tinker Bell. "What mischief is this, Tink?" he said.

"Is she dying?" asked Tootles.

"No," answered Peter, "fairy folk cannot be harmed by mortal hand. She's just lying there to get sympathy."

"Silly ass," said Tink, and flew up into the tree.

"You've not heard the last of this, Tink!" Peter said angrily. "You are my friend no more!"

Now as Peter moved out of the shadows of the tree in which Tinker Bell sulked into the full moonlight, Bertie saw that he was a most amazing boy, indeed. His head had hardly any hair, and was set high-off by his pointed, elongated ears. His eyes were slanted as if of an Oriental cast, and the nose was flat and simian. Only the mouth was like a normal boy's, petulant, and slightly parted, revealing the perfect pearls of his baby teeth.

Yet in his manner, Peter Pan was like a military commander. At his side, dangling down to his knees, was a long slender sword, now sheathed in a scabbard of ebony. He wore not clothes, but multi-colored leaves which seemed to have been stuck onto his body. He was so slender it was as if his body had reached the threshold of adolescence, and then rebelled against growing further. He seemed half-boy, half-fairy.

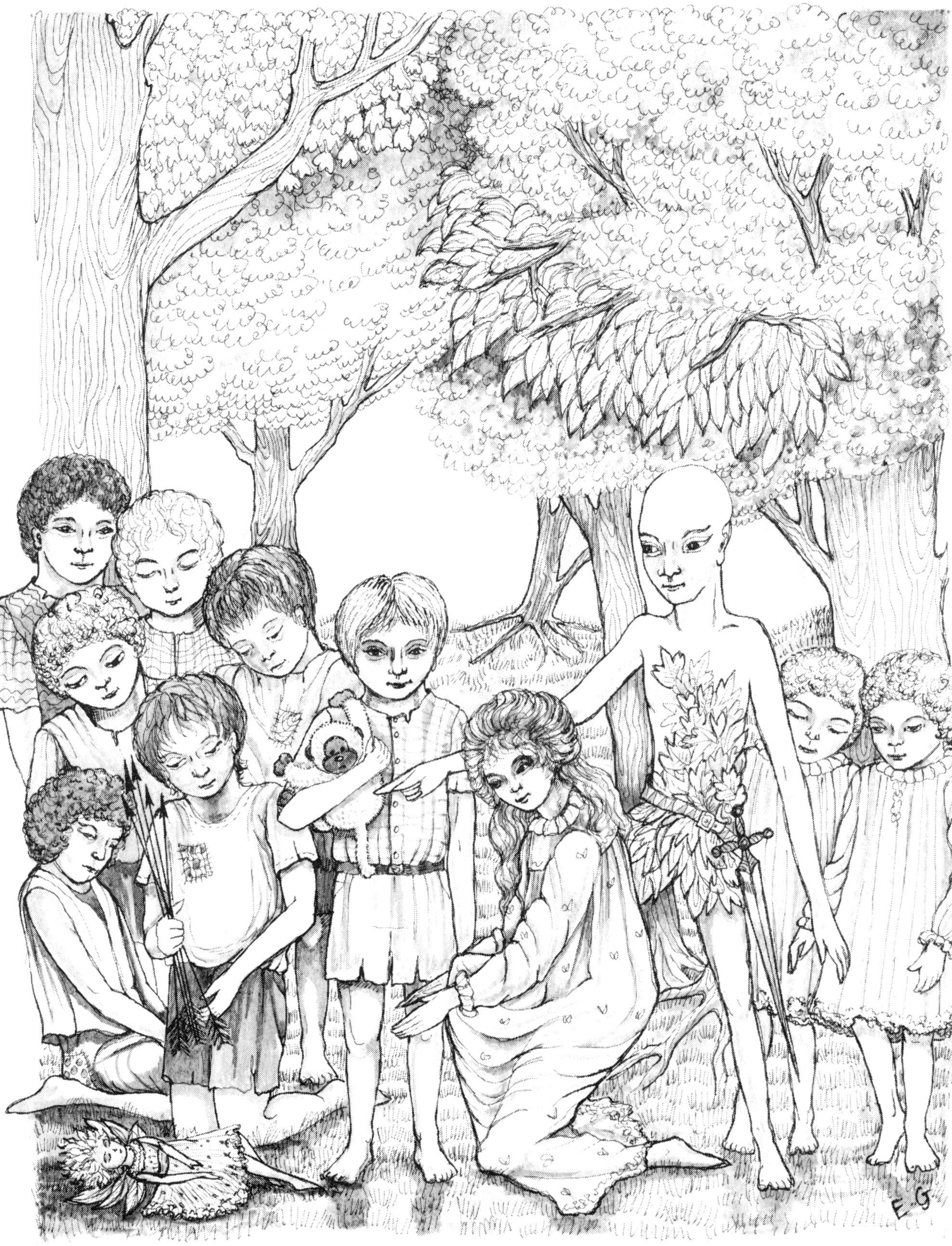
E.G

Now Peter barked commands at the Lost Boys. With him, besides Wendy, he had brought John and Michael, her two younger brothers. But there were the two unexpected arrivals as well, Bertie and Bongo.

"Nibs!" called Peter. "Run to Tiger-Lily and ask her for ten braves to guard our camp tonight. We shall have to sleep on the ground here, and tomorrow our guests can be measured for trees."

This was the secret of the Lost Boy's hidden home underground, a very delightful residence of which we shall see a good deal more presently. But there is no entrance to be seen, not so much as a pile of brushwood. Look closely, however, and you may note that there are here seven large trees—precisely the number of the Lost Boys plus Peter—each having in its hollow trunk a hole as large as a boy. These are the seven entrances to the home under the ground, for which Hook has been looking in vain.

Presently the Redskins arrive. They are in Peter's debt because many moons ago, Peter had saved the life of Tiger Lily when Hook had plans to drown her on Marooner's Rock. The tale is worth the telling because it demonstrates the extent of Peter's fey powers.

The place, as I have said, is Marooner's Rock. The pirates had caught Tiger Lily boarding the ship with a knife in her mouth. No watch was kept on the ship, it being Hook's boast that the wind of his name guarded the ship for miles around. Now news of Tiger Lily's fate would enhance the invisible guard of Hook's evil name.

She was bound hand and foot and set in the stern of the pirate dingy. Starkey rowed and Smee guided him to Marooner's Rock.

"Luff, you lubber!" cried Smee. "Here's the rock. Now then, our business is to hoist the redskin on to it, and leave her here to drown."

Quite near the rock, bobbing up and down in the water was Peter. He saw the fate intended for the Indian princess, and could easily have foiled that fate, but he chose another way deliberately designed to raise Hook's ire. There was almost nothing he could not do, and now he imitated perfectly Hook's voice and inflections.

"Ahoy there, you lubbers!" he called.

"The captain," said the pirates, staring at each other in surprise.

"Set the redskin free," came the order.

"Free?!" cried Smee.

"Yes, cut her bonds and let her go."

"But, captain—"

"At once, d'ye hear me, or I'll pluck out your backbone with my hook!"

"This is queer," said Smee.

"Better do what the captain orders," said Starkey nervously.

"Aye, aye," Smee agreed, and he cut Tiger Lily's cords. At once like on eel she slid out of the boat and under the waves.

Well may we imagine Hook's rage, but when Smee and Starkey swore on his hook that they had heard his voice across the water issuing the orders to set her free, he ceased to blame his men and knew somehow Pan was behind the trick. Once more he vowed a terrible revenge.

For her part, Tiger Lily committed all the forces of her island Indian nation in the defense of Peter and the Lost Boys. And so it was that that night they slept soundly above ground, guarded by Tiger Lily's ten bravest braves.

The next day, at the first light of dawn, Peter was up issuing directions for the fitting of trees to the five newcomers. It was decided that Bongo would not need his own tree, being small enough to use any of those now available. The trees were hollow, and dropped into place so that there was a vertical passage to the home under the ground.

First you were measured, and then a tree hollowed out to fit exactly, for it had to be just the right size or you could not go up and down. You drew in your breath at the top, and down you went until your feet touched the dirt floor and woven grasses that served as carpets. To ascend, you wriggled up like a snake, pressing the soles of your feet against the sides of the hollow trunk.

After a few days practice they could all go up and down as gaily as buckets in a well. And how ardently they grew to love their home under the ground, especially Wendy because the role of mother to them all suited her perfectly.

Their home consisted of one large room. From the dirt floor grew stout mushrooms of a charming color which were used as stools. A Never tree tried hard to grow in the center of the room, but every morning they sawed the trunk through, level with the floor. By tea-time it was always about two feet high, and then they put a door on top, thus becoming a table, which when cleared and removed, was sawed again level with the floor in order to make more room to play. There was an enormous fireplace across which Wendy stretched strings from which she suspended her washing. The bed was tilted against the wall by day, and let down at night, and all the boys slept in it like sardines in a can. But Wendy would have a baby, and Bongo being the littlest, and you know how women are, so the long and short of it was that he was hung up in a basket.

It was rough and simple, and not unlike what baby bears might live in. But there was one mysterious recess in the wall, no larger than a bird-cage, which was the private apartment of Tinker Bell, who had been forgiven by Peter. It could be shut off from the rest of the home by a tiny curtain, which Tink (who like most fairies was quite fastidious) always kept drawn when dressing or undressing.

No woman, regardless of size, could have had a more exquisite boudoir and bedchamber combined. The couch, as she always called it, was a genuine Queen Mab, with club legs; and she varied the bedspreads according to what fruit-blossom was in season. Her mirror was a Puss-in-boots, of which there are now only three, unchipped, known to fairy dealers; the wash-stand was Pie-crust and reversible, the chest of drawers an authentic Charming the Sixth, and she bathed in a golden thimble, inscribed with the name MARIAN, which she claimed had been given to her by Maid Marian of Robin Hood fame, fairies being ageless, as we know. Tink was very contemptuous of the rest of the house, as was perhaps unavoidable, for they were indeed a motley crew. Her chamber, though beautiful, looked rather conceited, having the appearance of a nose permanently turned up.

I suppose it was all especially entrancing to Wendy, because those rampageous boys of hers gave her so much to do. Really there were whole weeks when, except perhaps with a stocking in the evening, she was never above ground. The cooking, I can tell you, kept her nose to the pot. Wendy's favorite time for sewing and darning was after they had all gone to bed. Then, as she expressed it, she had a breathing time for herself, and she occupied it in making new things for them, and putting double pieces on the knees, for they were all most frightfully hard on their knees.

Since some of you have only recently joined our story, swimming in with Bertie and Bongo so-to-speak, it is perhaps only fair that I tell you how Wendy, Michael, and John came to be flying to Neverland on the night that Bertie and Bongo arrived.

Wendy, John, and Michael were the children of Mr. and Mrs. George Darling, and lived at number 14, in Kensington, London. There never was a simpler, happier family until the coming of Peter Pan.

Mrs. Darling first heard of Peter when she was tidying up her children's minds. It is the nightly custom of every good mother after her children are asleep to rummage in their minds and put things straight for the next morning, replacing into their proper places the many articles that have wandered during the day. If you could keep awake (but of course you can't) you would see your own mother doing this, and you would find it very interesting to watch her. It is quite like tidying up drawers. You would see her on her knees, I expect, lingering humorously over some of your contents, wondering where on earth you had picked this thing up, making discoveries sweet and not so sweet, pressing this to her cheek as if it were as nice as a kitten, and hurriedly stowing that out of sight. When you wake in the morning, the naughtiness and evil passions with which you went to bed have been folded up small and placed at the bottom of your mind; and on the top, beautifully aired, are spread out your prettier thoughts, ready for you to put on.

I don't know whether you have ever seen a map of a person's mind. Doctors sometimes draw maps of other parts of you, and your own map can become intensely interesting, but catch them trying to draw a map of a child's mind, which is not only confused, but keeps going round all the time. There are zigzag lines on it, just like your temperature on a card, and these are probably roads in the island; for the Neverland is always more or less an island, with astonishing splashes of color here and there, and coral reefs and rakish-looking craft in the offing, and savages and lonely lairs, and gnomes who are mostly tailors, and caves through which a river runs, and princes with six elder brothers, and a hut fast going to decay, and one very small old lady with a hooked nose.

It would be an easy map if that were all; but there is also your first day at school, religion, fathers, the round pond, needlework, verbs that take the dative, fractions, Show and Tell, your favorite flavor ice-cream, slivers and iodine, leaving a molar under your pillow for the tooth-fairy, Christmas, braces on your teeth, innoculations, and so on; and either these are part of the island or they are another map showing through, and it is all rather confusing, especially as nothing will stand still.

Of course the Neverlands vary a good deal. John's, for instance, had a lagoon with flamingoes flying over it at which John was shooting, while Michael, who was very small, had a flamingo with lagoons flying over it. Of all the delectable islands, the Neverland is the snuggest and most compact; not large and sprawly, you know, with tedious distances between one adventure and another, but nicely crammed. When you play at it by day, with chairs and table-cloth, it is not in the least alarming, but in the two minutes before you go to sleep it becomes very nearly real. That is why there are night-lights.

Occasionally in her travels through her children's minds Mrs. Darling found things she could not understand, and of these quite the most perplexing was the word Peter. She knew of no Peter, and yet he was here and there in John and Michael's minds, while Wendy's began to be scrawled all over with him. The name stood out in bolder letters than any of the other words, and as Mrs. Darling gazed she felt that it had an oddly cocky appearance.

"Yes, he is rather cocky," Wendy admitted when her mother questioned her.

"But who is he, my pet?"

"He is Peter Pan, you know, mother."

At first Mrs. Darling did not know, but after thinking back into her childhood she just remembered a Peter Pan who was said to live with the fairies. There were odd stories about him; as that when children died he want part of the way with them, so that they should not be frightened.

One night some leaves of a tree had been found on the nursery floor, which certainly were not there when the children went to bed, and Mrs. Darling was puzzling over them when Wendy said with a tolerant smile:

"I do believe it is that Peter again."

We now come to the night on which the extraordinary adventures of these children may be said to have begun. It happened to be Nana's evening off. Nana was the Darling's nursemaid, but she was no ordinary nursemaid. No, indeed. In fact she was a prim Newfoundland dog, who had belonged to no one in particular until the Darlings engaged her. She had always thought children important, however, and the Darlings had become acquainted with her in Kensington Gardens, where she spent most of her spare time peeping into perambulators, and was much hated by careless nursemaids whom she followed to their homes and complained to their mistresses. She proved to be quite a treasure of a nurse. How thorough she was at bath-time; and up at any moment of the night if one of her charges made the slightest cry. Of course her kennel was in the nursery. She had a genius for knowing when a cough is a thing to have no patience with and when it needs a stocking round your throat. She believed to her last day in old-fashioned remedies like rhubarb leaf, and made sounds of contempt over all this new-fangled talk about germs, and so on. No nursery could possible have been conducted more correctly, and Mr. Darling knew it, yet he sometimes wondered uneasily whether the neighbors talked.

Anyhow, this was Nana's night off, and Mrs. Darling had bathed and sung to the children till one by one they had let go her hand and slid away into the land of sleep. All were looking so safe and cosy that she smiled and sat down tranquilly by the fire to sew. Then her head nodded, oh, so gracefully, and she was asleep.

While she slept she had a dream. She dreamt that the Neverland had come too near and that a strange boy had broken through from it. He did not alarm her, for she thought she had seen him before in the faces of many women who have no children. Perhaps he is to be found in the faces of some mothers also. But in her dream he had rent the film that obscures the Neverland, and she saw Wendy and John and Michael peeping through the gap.

The dream by itself would have been a trifle, but while she was dreaming the window of the nursery blew open, and a boy did drop onto the floor. He was accompanied by a strange light, no bigger than your fist, which darted about the room like a living thing, and I think it must have been the light which awakened Mrs. Darling.

She started up with a cry, and saw the boy, and somehow she knew at once that he was Peter Pan. If you or I or Wendy had been there we should have seen that he was very like Mrs. Darling's kiss. He was a lovely boy, clad in skeleton leaves and the juices that ooze out of trees; but the most entrancing thing about him was that he had all his first teeth. When he saw she was a grownup, he gnashed the little pearls at her.

Mrs. Darling screamed, and, as if in answer to a bell, the door opened, and Nana entered, returned from her evening out. She growled and sprang at the boy who leapt lightly through the window. She looked out, but could see nothing but what she thought was a shooting star.

She returned to the nursery, and found Nana with something in her mouth, which proved to be the boy's shadow.

A week later, on a day the Darling household was to refer to ever after as "that never-to-be-forgotten Friday," Peter returned with Tinker Bell for his shadow.

The Darlings had gone to number 27 for dinner, leaving Nana chained in the backyard due to a misunderstanding over a joke Mr. Darling had played upon her by filling her milk bowl with his own horrid medicine. The joke had backfired upon him because all the family had sympathized with Nana, and to show his authority he had ordered her tied up in the yard. No sooner had the door of 27 closed on Mr. and Mrs. Darling than there was a commotion in the stars overhead, and the smallest of all the stars in the Milky Way screamed out: "Now Peter!"

For a moment after Mr. and Mrs. Darling left the house the night-lights by the beds of the three children continued to burn clearly. They were awfully nice little night-lights, and one cannot help wishing that they could have kept awake to see Peter; but Wendy's light blinked and gave such a yawn that the other two yawned also, and before they could close their mouths all the three went out.

There was another light in the room now, a thousand times brighter than the night-lights, and in the time we have taken to say this, it has been in all the drawers of the nursery, looking for Peter's shadow. It made this light by flashing about so quickly, but when it came to rest for a second you saw it was a fairy, no longer than your hand, but still growing, called Tinker Bell.

A moment after the fairy's entrance the window was blown open by the breathing of the little stars, and Peter dropped in. He had carried Tinker Bell part of the way, and his hand was still messy with the fairy dust.

Nana knew at once something was wrong and began a furious barking in the backyard. Wendy was now awake and in deep conversation with Peter.

"I don't want ever to be a man," he said with passion. "I always want to be free to have fun and adventures. To fly is to be free. Oh, Wendy, I could show you how!"

"I ran away the day I was born because I heard my mother and father talking about what I was to be when I became a man. So I ran away to Kensington Gardens and lived a long time among the fairies."

"But where do you live now?" Wendy asked.

"In Neverland with the lost boys."

"Who?"

"They are the children who fall out of their perambulators when the nurse is looking the other way. If they are not claimed in seven days they are sent far away to Neverland to defray expenses. I am their captain," and as he said this, the cocky little fellow crowed like a chicken.

"What fun it must be!"

"Yes," said the cunning Peter, "but we are rather lonely. You see we have no female companionship."

"Are none of the others girls?"

"Oh no, girls, you know, are much too clever to fall out of their prams."

This flattered Wendy immensely. "I think," she said, "that it is perfectly lovely the way you talk about girls."

"Wendy," said the sly one, "you could tuck us in at night."

"Oo!"

"None of us has ever been tucked in at night."

"Oo," and her arms went out to him. In no time she was giving him thimbles for kisses, and kisses for thimbles, a confusion that they soon worked out because he did not know the meaning of the word kiss.

But when Peter thimbled her, it felt exactly like someone pulling her hair, and, indeed, it was the jealous Tink.

"She says she will do that to you, Wendy, every time I give you a thimble."

Then Peter changed tactics once more and appealed to Wendy's mother instinct.

"You could tell us stories after tucking us in."

How could she resist? "Of course its awfully fascinating!" she cried. "Would you teach John and Michael to fly too?"

"If you like," he said indifferently; and she ran to John and Michael and shook them. "Wake up!" she cried. "Peter Pan has come to teach us to fly!"

In the yard below, seeing that no help would come from her barking, Nana had strained and strained at the chain until it broke. In another moment she had burst into the dining room of 27 and flung up her paws to heaven. Mr. and Mrs. Darling knew at once something terrible was happening in their nursery, and without a goodbye to their hostess they rushed into the street.

Peter had blown fairy dust on the children, as subtly as blowing a kiss, and from below as they hurried down the street, Mr. and Mrs. Darling could see the bedroom ablaze with light, and four little figures in night attire circling round and round, not on the floor but in the air.

In a tremble they opened the street door. Will they reach the nursery in time? If so, how delightful for them, and we shall all breathe a sigh of relief, but there will be no story. On the other hand, if they are not in time, I solemnly promise that it will all come right in the end.

They would have reached the nursery in time had it not been that the little stars were watching them. Once again the stars blew the window open, and that smallest star of all—the wicked Tinker Bell called out:

"Out, Peter!"

Then Peter knew that there was not a moment to lose. "Come!" he cried, like the captain of the lost boys, and soared out into the night followed by John, Michael, and Wendy. Tinker Bell lit the way straight on till morning.

Mr. and Mrs. Darling rushed into the nursery too late. The birds were flown.

And so the happy domestic scene we see in the house underground, the lost boys all tucked in and storied too, came at some expense to another mother and father. Laying the blame on the caper that had caused him to tether Nana in the yard for the loss of his children, Mr. Darling took to living in Nana's doghouse in the nursery and nothing could coax him out.

For her part Wendy took to pretending to be the mother of the lost boys. "You see, I feel that is exactly what I am."

"It is, it is!" they all cried. "We saw it at once."

Before Wendy's coming, the subject of mothers was one of Peter's taboos, because of possible desertions in his ranks if the call of home became too great. Now, by presenting the boys with a proper substitute, they could have all the mothering they needed.

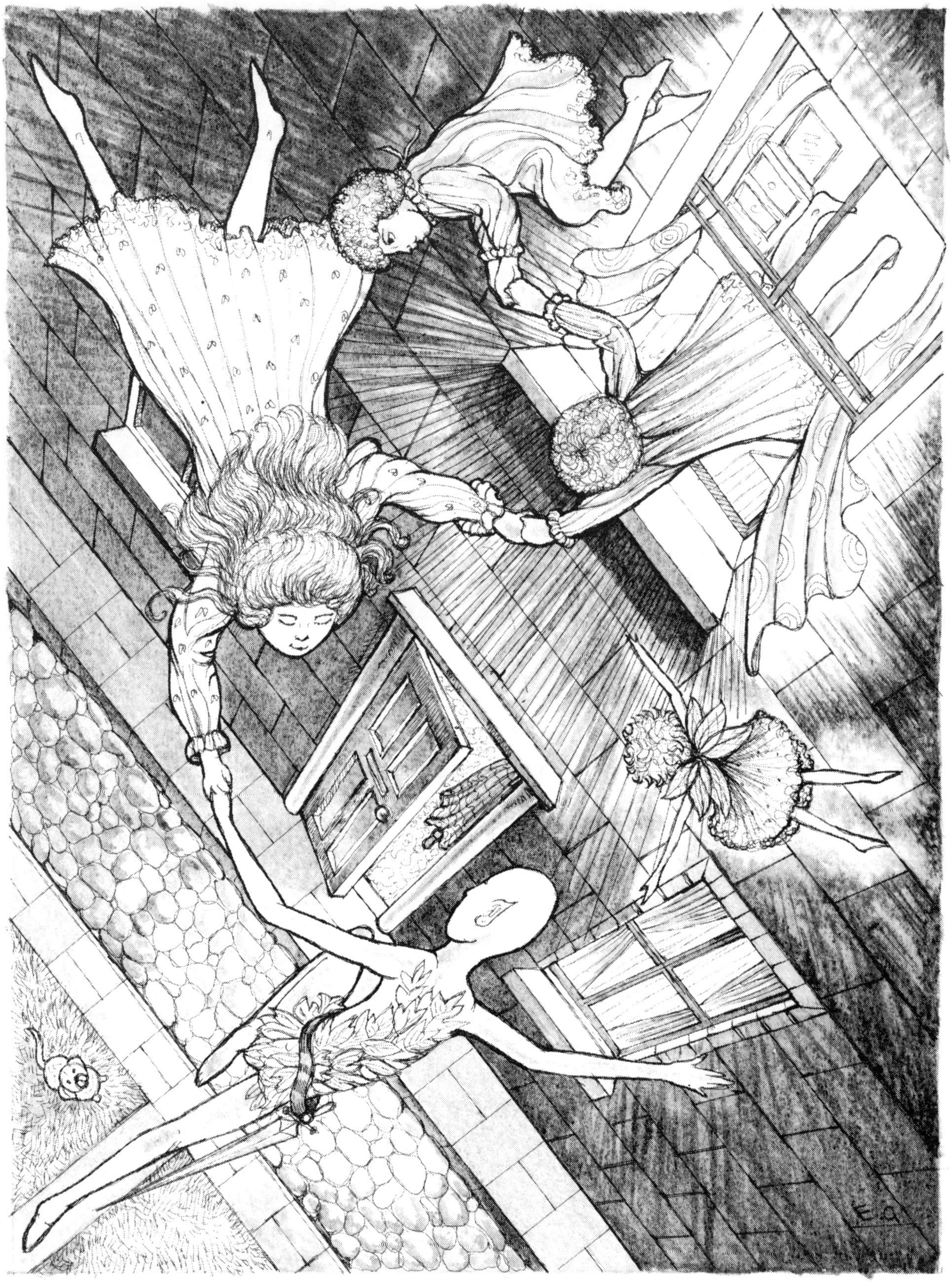

Talking about real mothers, however was still taboo. Peter saw them as a restrictive influence, possibly limiting the desire for adventure and Peter's own aggressive feelings towards Hook: "Hook or me!" Imagine what it would be like, he thought, if every boy's mother was running around the island. They would never be able to get on with the business with the pirates.

But it was Peter's attitude towards mothers that brought him in conflict with the newcomer Bertie. Bertie reminded Peter that he was *not* a lost boy, and, therefore, he saw no reason why he should honor Peter's taboos or respect his captaincy.

Bertie had a purpose, which he spoke about from time to time, which was tied-up with his mother as tightly as an umbilical cord. And, so, he told Peter, he would darn well talk of *his* mother whenever it pleased him. So flushed with victory was Peter at adding two more lost boys to his ranks, *plus* a pretend mother, he was content for now to overlook this minor insurrection in the ranks.

But all was not bliss for Wendy. Tink's jealously never let up. She could not do much to Wendy when Peter was around, but at night when Wendy did the mending and all the others slept, she tormented Wendy by flying into her elbow, causing the needle to prick her other thumb. By the end of her first week, she could no longer go on. Peter demanded to know why, since mending took second place in the order of mothering only to telling stories at bedtime.

And so Wendy had to tell on Tink for her own preservation. Peter was so infuriated this time that he went to Tinker Bell's tiny niche in the wall, ripped the curtain aside, and plucked out the golden thimble. Now this was Tinker Bell's most revered heirloom because it had been given to her by Maid Marian, no less than Robin Hood's own betrothed. It served Tink as a bath, and as we have seen with the fairy Mrs. B., cleanliness is of the utmost priority with fairies.

But Peter told Tink she could bathe in the lagoon and *gave* the thimble to Wendy to protect her ravaged thumb against the further predations of Tinker Bell. At first Tinker Bell raged, then cried enough tears to fill the thimble, but Peter would not relent. From then on, she stayed away from Wendy at night, but plotted her revenge.

Bertie came into conflict again with Peter when he expressed his doubts to the other boys that Tink was actually a fairy. After all, he had seen only *good* fairies before.

"I know a big fairy that would cure Tinker Bell of pinching people," said Bertie to the lost boys who loved to sit in a circle and hear of his adventures when Peter was away on his own mysterious adventures. "She'd put Tink in a place where she'd be pinched all day and all night, and that would cure her for good, you bet."

But although the Lost Boys liked to hear tell of Bertie's life under the sea in *The Water-Babies*, its morality was beyond them, or rather they were in that pre-adolescent stage in which they behaved—without the guidance of mother or father—like little savages.

Peter set the example for them, of course. Being utterly free and irresponsible, flying wherever he wished, Neverland, with its blend of adventure, aggression, and fantasy suited him perfectly. If he had an overwhelming desire or wish, it was "Hook or me," the final encounter with Hook, sword play to the death.

And as Peter's infernal cockiness caught in Hook's craw, the great pirate who had destroyed all his rivals on the Spanish Main had for himself now only one goal, the annihilation of Peter Pan. And so like two stars orbiting one another, Hook all black, and Peter effervescent and airy, with Tinkerbell's light always around him, they circled one another in Neverland.

We come now to that evening which was to be known among them as the Night of Nights, because of its adventures and their upshot. The day, as if quietly gathering its forces, had been almost uneventful, and now the redskins in their blankets were at their sentry posts above, while, below, the children were having their evening meal.

This meal happened to be a make-believe tea, and they sat round the board, guzzling in their greed; and really, what with their chatter and recriminations, the noise, as Wendy said, was positively deafening. To be sure, she did not mind noise, but she simply would not have them grabbing things, and then excusing themselves by saying that Tootles had pushed their elbow. There was a fixed rule that they must never hit back at meals, but should refer the matter of dispute to Wendy by raising the right arm politely and saying, "I complain of So-and-so," but what usually happened was that they forgot to do this or did it too much.

"Silence!" cried Wendy when for the twentieth time she had told them that they were not all to speak at once. "Is your calabash empty, Slightly, darling?" Naturally on the island they had not plates and cups, but instead used hollowed-out gourds and squashes.

"Not quite empty, Mummy," Slightly said.

"He hasn't even begun to drink his tea," Nibs interposed.

Under the complex set of rules that governed their behavior, this was "telling," and Slightly seized his chance.

"I complain of Nibs," he said promptly.

John, however, had held up his hand first.

"Well, John?"

"May I sit in Peter's chair, as he is not here?"

"Sit in father's chair, John!" Wendy was scandalized. "Certainly not.

"He is not really our father," John answered. "He didn't even know how a father behaves until I showed him how."

This was grumbling. "We complain of John," cried the Twins.

Tootles held up his hand. He was so much the humblest of them, indeed the *only* humble one, that Wendy was especially gentle with him.

"I don't suppose," Tootles said diffidently, "that I could be father."

"No, Tootles."

Once Tootles began, which was not very often, he had a silly way of going on.

"As I can't be father," he said heavily, "I don't suppose you would let me be baby."

"Now you *know* Bongo is baby," Wendy reproved him. He's the smallest and it's only fair."

From his hanging basket, which stood for a cradle, Bongo looked over at Tootles as if to say, "Hard cheese, mate." Then, as if to further exploit his privileged status, he picked that moment to raise his milk bottle to his lips and began smacking them in delight. This was particularly galling to the boys because Bongo was the only one to have the luxury of milk, there being only one cow on the island. Peter openly scorned milk as baby food, and so the boys had to follow his example, but there is nothing, nay, even candy, that a young boy misses more than a cool glass of milk.

And so, as if in chorus, they all said at once, "I complain of Bongo."

"Do try not to smack your lips so, dear Bongo," said Wendy.

Bertie went up to Bongo's basket and began observing him closely. Slightly persisted.

"As I can't be baby," Tootles said getting heavier and heavier, "do you think I could be a twin, Mummy?"

"No, indeed," replied the twins, "it's frightfully difficult being a twin."

"As I can't be anything important," said Tootles, "would any of you like to see me do a magic trick?"

"No," they all replied, for Tootles tricks fooled only himself.

Then at last he stopped. "I hadn't really any hope," he said, and sat down.

"That's funny," said Bertie, watching Bongo drink his bottle, "I could have sworn he didn't have a mouth, and now he does."

"Perhaps he grew one," laughed Nibs.

Bongo winked at Bertie. Then the hateful telling broke out in a rash of accusations.

"Nibs is speaking with his mouth full."

"Slightly is coughing on the table."

"I complain of the Twins."

"I complain of Curly."

"I complain of John and Michael."

"I complain of Bertie and Bongo."

"Oh dear, oh dear," cried Wendy. "I'm sure I sometimes think that children are more trouble than they are worth."

She told them to clear away the table, and sat down by the fire with her workbasket: a heavy load of stockings and every knee with a hole in it as usual. Was it a trick of the fire that made her hair appear grey?

While she sewed they played around her; such a group of happy faces and dancing limbs lit up by that romantic fire. This had become a very familiar scene for many weeks in the underground home, but we are looking upon it for the very last time. Peter entered from his tree.

"Ah, old lady," Peter said to Wendy, looking down at her as she sat mending a heel, "there is nothing more pleasant of an evening for you and me when the day's toil is over than to rest by the fire with the little ones near by."

"It is sweet, Peter, isn't it?" Wendy said frightfully gratified. "Peter, I think Bongo has your nose."

"Michael takes after you."

Now came their favorite time, when she told her nightly story. It was always the same story, and they all knew it by heart, but they pretended not to know.

"Listen, then," said Wendy, settling down to her story. "There was once a gentleman—"

"I had rather he had been a lady," interrupted Curly.

"I wish he had been a white rat," said Nibs.

"Quiet," their mother admonished them. "There was a lady also—"

"O Mummy," cried the first twin, "you mean that there is a lady also, don't you? She is not dead, is she?"

"Oh, no."

"I am awfully glad she isn't dead," said Tootles. "Are you glad, John?"

"Of course I am."

"Are you glad, Nibs?"

"Rather."

"Are you glad, Twins?"

"We are just glad."

"How about you, Bertie?"

Bertie said nothing. Because of his purpose, Bertie had trouble pretending Wendy was his mother.

"Little less noise there!" Peter called out, determined that she should have fair play, however beastly a story it might be in his opinion.

"The gentleman's name," Wendy continued, "was Mr. Darling, and her name was Mrs. Darling."

"I knew them," John said, to annoy the others.

"I think I knew them," said Michael rather doubtfully.

"They were married, you know," explained Wendy, "and what do you think they had?"

"White rats," cried Nibs, inspired.

"No."

"It's awfully puzzling," said Tootles.

"Quiet, Tootles. They had three descendants."

"What is descendants?"

"Well, you are one, Twin."

"Do you hear that, John? I am a descendant."

"Descendants are only children," said John.

"Oh, dear, oh dear," sighed Wendy. "Now these three children had a faithful nurse called Nana; but Mr. Darling was angry with her and chained her up in the yard; and so all the children flew away."

"It's an awfully good story," said Nibs.

"They flew away," Wendy continued, "to the Neverland, where the lost children are."

"I just thought they did," Curly broke in excitedly. "I don't know how it is, but I just thought they did."

"O Wendy," cried Tootles, "was one of the lost children called Tootles?"

"Yes, he was."

"I am in a story. Hurrah, I am in a story, Nibs."

"Hush. Now, I want you to consider the feelings of the unhappy parents with all their children flown away. Think of the empty places at the table, and the downy pillows that miss the children's cheeks."

"Oooo!" They all moaned, though they were not really considering the feelings of the unhappy parents one jot.

"It's awfully sad," the first twin said cheerfully.

"I don't see how it can have a happy ending," said the second twin. "Do you, Nibs?"

"I'm frightfully anxious."

"If you knew how great is a mother's love," Wendy told them triumphantly, "you would have no fear. You see, our heroine knew that the mother would always leave the window open for her children to fly back in; so they stayed away for years and years and had a lovely time."

"Did they ever go back?" someone muttered.

"Let us now," said Wendy, bracing herself for her finest effect, "take a peep into the future. See, dear brothers," says Wendy, pointing upwards, "there is the window still standing open, and now we are rewarded for our sublime faith in a mother's love, for who are the three figures who fly up to be embraced by their mother and father?"

"Am I one?" said Michael uncertainly, who always asked this question at this point.

"Yes, Michael."

"Me, too, Wendy," said John.

"Yes."

This was the part of the story that Peter hated most. Tonight he could contain himself no longer.

"You are wrong, Wendy, about mothers."

"Oh," she said, surprised.

"Long ago," he said, "I thought like you that my mother would always keep the window open for me; so I stayed away for moons and moons and moons, and then flew back, but the window was barred, for mother had forgotten all about me, and there was another little boy sleeping in my bed."

I am not sure that this was true, but Peter thought it was true; and it scared them.

"Are you sure mothers are like that?" Tootles asked.

"Yes," said Peter, "positively."

So this was the truth about mothers. The toads!

"You lie!" shouted Bertie. No one had ever questioned Peter's authority before, and they all turned aghast to look at him, then back again to Peter for his reaction. Peter said nothing, but his look betrayed his dark mood.

"*My* mother would leave the window open for me."

"Then why don't you go back and find out for yourself!" shouted Peter.

"That's very cruel of you," said Wendy. "You know Bertie is an orphan."

"I *will* go back!" cried Bertie, "and I'll rescue my mother from time. And you'd do the same thing too, Peter, if you were a man."

"I'm not a man, and never want to be!" exclaimed Peter.

"Then you'll rot in Neverland. Look at that puny sword of yours. All you can do is kill *pretend* pirates."

"I cut off Hook's arm with it," said Peter, "but I wish it had been something else."

"He means his *head*," said Nibs.

"You probably don't like fathers anymore than mothers, do you Peter?" said Bertie pressing his argument. "In fact you don't like grown-ups at all."

This kind of insurrection in the ranks was more than Peter could take, and with that he angrily left. Had his tree a door, he would have slammed it.

Peter's quitting the field to Bertie had an odd effect upon the other boys. For the moment, at least, Bertie had become number one, and the boys crowded around him.

"Tell us again what Toad said to the washerwoman," said Nibs.

"Oh, no, no! The picnic, the picnic!" cried John.

"The badger, the badger!" called Curly.

"Toad crowed just like Peter," recalled Bertie. "He was always making up songs about his greatness." Quoting toad, " 'Who was it had all of them snowed?'" Striking an heroic pose, "'None other than the magnificent, all-glorious Toad.'"

Bertie's recital of Toad's follies would set the boys to laughing and clapping, and it always got Bongo to holding his sides and rolling on the ground, which made the boys laugh even harder.

"Too bad Toad could not fly like Peter," said Nibs in defense of his leader.

"Oh, but he did," said Bertie. "After everyone thought he was reformed, he got a plane and buzzed Toad Hall."

From their blank looks, Bertie realized they'd never, ever heard about airplanes.

"Haven't you heard about the Wright Brothers?"

More blank looks.

"What about Kitty Hawk?"

"Who's she?" asked Tootles.

153

"Oh, it's not a she," replied Bertie, "it's a place in North Carolina in the United States where Orville and Wilbur Wright made the first plane for flying through the air."

"That's not so hard," said Curly. "We've all made paper planes that fly through the air.

"These aren't paper!" shouted Bertie. "The Wright Brother's plane was made out of wood so that one of them could sit in it and take off and land with a *motor*!"

"Pooh-pooh," said someone, and they were all laughing at him. "Must have been the Wrong Brothers," joked Nibs.

"Now be fair," said Wendy. "Hear him out."

"I see what the trouble is," said Bertie. " *Your* time comes before the Wright Brothers. The airplane hasn't been invented yet."

No matter what he said, the boys still laughed and carried on. Then a bright idea came to him.

"Who was queen of England when you left home with Peter?" he asked, looking over to John and Michael.

"I don't know her name," said Michael, "but I think she is a nice lady."

"Victoria," said Wendy, answering Bertie's question.

"Well that's it then," said Bertie. "*Your* time in England comes before my time, and the airplane isn't invented until 1903."

Now everyone was really excited. To them Bertie was like a time-traveller, bringing news of their own futures. Of course, to boys there is nothing more exciting than warfare, whether in the form of Lost Boys and pirates, cops and robbers, cowboys and Indians, or soldiers and sailors of warring countries. They play this game as naturally as girls play with dolls. Can there be hope then for the future? Let us pray that the boys of the world grow up in time before the lost boys of Neverland.

But now they crowded around Bertie, full of questions about the mysterious future.

"How far can these planes fly?" queried John.

"Oh, *thousands* of miles," replied Bertie.

"How *fast*?" inquired Nibs.

"Well," said Bertie, "that all depends on what kind of plane it is. The fighters have to shoot down the bombers, so they have to be faster—three to four hundred miles an hour."

"Four hundred miles an hour!" they cried out ecstatically.

Nibs held out his arms like wings and circled around the room.

"I'd want to be a fighter," said Nibs, "and shoot down bombers."

"No, a fighter is the type of plane," corrected Bertie. "The man who flies it is a pilot."

"Is there room for *two*?" asked the Twins.

"Oh, yes," responded Bertie, "some of the bombers have four huge motors and carry ten to twelve crew men."

"What do the bombers do?" asked Tootles, naively.

His question brought a storm of ridicule about his head. When the commotion had died down again, Curly had yet another question for Bertie: "Do the crew men throw the bombs out, and what are they supposed to hit?"

The boys all gleefully made motions as if throwing bombs, but Bertie interrupted their sport.

"No," he said, "some of the bombs weigh as much as a ton—"

"Wow!" came the cry.

"—So they have to drop them mechanically. That's the bombardier's job, and he has a bombsight so that he can drop the bombs with pinpoint accuracy."

"Wow!" came the collective cry again.

"I want to be the bombardier," exclaimed Curly.

"No, *me*!" shouted Michael. And then they were all calling out, "Me!Me!Me!" until Wendy had to stop her ears with her hands and threaten to send them all off to bed.

"How high do they fly?" asked one of the Twins.

"Up to fifty-thousand feet," answered Bertie, "and the crew has to breathe oxygen from tanks or they would die."

"Ten miles high!" cried John, showing off his prowess at math.

"And they can hit the point of a pin from there?" questioned the other Twin incredulously.

"Oh, no, 'pin-point accuracy' is just an expression," said Bertie smiling.

For the moment they had run out of questions. Slightly, who played pipes like Peter, and who had not been heard from, now threw himself into the conversational breach. "Does her Majesty still have a Navy?" he wanted to know.

"The greatest in all the world!" exclaimed Bertie.

"Then they'll still need bosuns," he replied, and played on his pipes the tune for piping someone aboard ship.

"Do they have any wars to use these planes in?" asked Nibs hopefully.

"Oh, yes," said Bertie, "two *world* wars!"

"Golly, gosh, and gee!" came the cries from the boys.

"Do you mean," questioned Tootles, "that another world comes out of the sky to fight our world?" His question set off another commotion of ridicule.

"No, no, Tootles," answered Bertie when it was quiet again. "I mean that twice between your time and my time practically the whole world fights."

"Golleee!" they cried.

"What are some of the planes named?" John asked.

"Well, there are dozens and dozens of different kinds," Bertie replied, "but the best fighter plane of all—I think—was called the Spitfire, because you see the machine guns are in the wings, and when they shoot it looks like it's spitting fire."

"Wow!" said the Twins as one.

155

"But then again," reflected Bertie, "the Americans believed the best fighter plane was their Mustang."

This remark set off a mock dogfight among the boys, culminating in their pinching one another while calling out, "Mus-sting, mus-sting!"

"No, no," said Bertie, "Mustang, Mus-tang. The name is a breed of wild horse."

"Why on earth would anyone name a plane for a horse," John inquired indignantly. "Her majesty should appoint a wartime committee to name planes, and I'd be the first to apply for the job."

"No, me, me!" shouted Michael.

"What are some of the bombers called?" asked the Twins as if of one mind.

"Sterlings, and Halifaxes, and Wellingtons," but that was as far as Bertie got.

"Wellingtons!" cried Nibs, who usually slept through history class. "What naming a plane after an old pair of 'Wellie' boots."

"No, you twit," said John, "the *Duke* of Wellington who defeated Napoleon at Waterloo.

"Go on Bertie, tell us some more names," urged Slightly.

"Well, the biggest British bomber is called the Lancaster. The Americans have one named the Flying Fortress with guns all over the place, so that they can shoot down the German planes from any angle."

"Good grief," cried Tootles, "aren't the Germans our friends?"

"Not in the first world war, and not in the second," replied Bertie.

"But Queen Victoria's husband, Prince Albert, is a German," said John.

"Do they shoot him as a spy?" asked Tootles.

"How do we come to fight the Germans, and who else do we fight?" asked Curly.

"Oh, it's all so complicated and difficult to explain," replied Bertie. Then he saw that Bongo was motioning to him, waving a paw and pointing into the air, and he knew that Bongo wanted to play his favorite game, Red Baron, with him.

"I'll tell you," said Bertie, "I'll show you Bongo's favorite game."

Going to Bongo, Bertie took out of his pocket a small pair of goggles, which he fitted over Bongo's eyes. "During World War I the greatest flying ace—"

"What's an ace?" Nibs interrupted.

"—a pilot who shoots down many other planes," responded Bertie. "This German ace was called the Red Baron, and his squadron of planes was called The Flying Circus."

With these words, Bertie held the sitting Bongo in the palm of his hand, simulating a flight through the air, while Bongo looked below to the ground over which he "flew" as Bertie ran with him around the room, all the time imitating a plane's throbbing engine. Again and again, he passed over their cheering heads, until Bertie grew tired of the game and set Bongo on his knee.

"Please, could I ask one more question," inquired Tootles.

Bertie nodded.

"If we drop bombs on Germans, how do we find the soldiers from way up in the sky?" The boys were surprised at the sophistication of Tootles' question, and hushed respectfully, awaiting Bertie's answer.

Bertie thought a moment and then replied, "Well, mostly they try to drop the bombs on factories that make weapons, but sometimes they bomb whole cities."

"That's terrible!" said Wendy, who had kept her peace so far, because as their mother she had no enthusiasm for their war games.

"But people live in cities," whined Tootles.

"Yes, they do, Tootles, and millions of people will die in the first world war. Millions of Germans, and millions of British."

"Millions?!" someone said incredulously.

Immediately a change came over each one of them. They suddenly sickened and lost their appetite for war.

"Why doesn't God stop them?" asked Tootles, back from the dead.

No one said a word, ashamed, everyone looked at the floor of the happy home which would never be the same again. Bertie was now very sorry that he had undertaken the role of prophet for the future. Then, speaking very tentatively, John spoke up.

"Are we *trying* to hit factories and things like that and *not* people with *our* bombs?" he asked.

"Yes," said Bertie, "but near the end, when they are losing, the Germans are not. They have a bomb with wings which they just point towards London and it flies until its clock stops. Then the motor stops too, and it comes down, and blows up anyone where it falls."

"Just like the crocodile," said Slightly.

"How so?" Nibs asked him.

"Well, everyone knows to run from it when they hear the clock ticking, but when it stops—bang!—the crocodile will swallow someone."

Bertie began crying, not because of the crocodile, but because only now after telling about the flying bomb did he recall the one that had landed on his house.

"Look what you've done, Slightly," Wendy said annoyed.

"Yes, you've made him cry," said Tootles.

Bongo climbed up from Bertie's knee and put his arms around his neck, comforting him.

Treasure this moment in Neverland, boys, where time does not tick, and none grow old, for evil creeps nearer to the happy home, and none can escape it. They cannot know their fate, but we do.

"Captain" Nibs blown up by a shell at Verdun; Curly caught in barbed wire and machine-gunned at the Somme; Slightly torpedoed and drowned in the service of Her Majesty's Navy; and the Twins gassed in a trench in France. The flower of a generation, English, French, and German, fallen to earth, buried.

Only Tootles, turned down for service, survives. As in Neverland where the big things constantly eluded him, World War I missed him too. In his old age he often gazes into the fire on his hearth and goes back in his mind to the snug little home underground; hears Wendy telling of the eternally open window of mother love into which they would all one day fly; sees again all the happy faces of childhood flickering in the firelight.

"Hallo, Tootles," he hears Nibs call to him.

"Would you like to see me do a trick?" Tootles asks him.

"No," comes the answer as always.

The Twins: "We complain of Tootles, Mummy, he hasn't touched his tea."

Curly's voice: "Are you a redskin or boy today, Tootles?"

"Boy."

"Come on, Tootles, don't lag behind," Slightly's voice.

John: "We're home now."

Michael: "Would you like to come with us, Tootles?"

"No." Then, "Yes, yes, take me with you," and the tears course down his face.

Recalling their fates, one by one, thinking perhaps that his fate to survive is the cruellest of all, brooding upon them night after day for the rest of his life.

They are truly lost boys.

An eerie sound of Pan pipes from above floats down the trees, breaking the silence that lay like a serpent coiled within the happy home. The sound recalled to Bertie's mind the Piper at the Gates of Dawn, in *The Wind and the Willows*, whose notes had told him that every living creature was protected, and that all was ultimately well. How to reconcile that with the despair over his parents' death that now clouded his mind completely.

Peter drops down his tree. He is accustomed to having at least two boys come to greet him, but no one rises. A pall, like the death of innocence, has fallen over the happy home. Peter had gone out to get the correct time for Wendy. The way one got the time on the island was to find the crocodile, and then stay near him till the clock struck.

"Eight bells, Mother," Peter called to Wendy.

"Thank you, Father," she said, playing the game one last time.

"No more dancing, children?" he asked, noting the mournful tone of the heretofore happy home.

"No," they answered sheepishly.

When they were tucked safely into bed, the fire burning down to a soft glow, Wendy went to Peter and put her hand on his shoulder.

"Dear Peter," she said, "with such a large family I have now passed my best years, but you haven't grown tired of me, have you?"

"No, Wendy, but...." He looked at her uncomfortably, blinking like one not sure whether he is awake or asleep.

"Peter, what is it?"

"I was thinking," he said, a little scared. "It is only make-believe, isn't it, that I am their father, because it would make me seem so old to be their real father."

"But they are ours, Peter, yours and mine."

"But not really, Wendy?" he asked anxiously.

"Not if you don't wish it," she replied; and she distinctly heard his sigh of relief. "Peter," she asked, trying not to let her voice betray her, "what are your exact feelings for me?"

"Those of a devoted son, Wendy."

"Do you ever think, when you grow up, that you might want to marry me?"

Now Wendy has broken the rules and made the pretend game real, opening the last forbidden door in Pan's psyche. Peter flees like the night before the dawn.

"I don't want to ever grow up," he said. "I want to stay a boy forever."

"I thought so," she said. "Well, goodbye, I'm not going to be your mother forever."

With her mind made up, she moved swiftly now, her sense of responsibility to her younger brothers clear in her mind.

"Wake up, John, Michael! Wake up, we're going back."

"Going back where?" they said.

"Home. Have you forgotten it?"

They had.

"Out of bed at once," Wendy snapped. Imagine Wendy snapping. "Peter, will you make the necessary arrangements for our return home?"

"If you wish it," he replied as coolly as if she had asked him to pass the nuts.

Not so much as a sorry-to-lose-you between them! If she did not mind the parting, he was going to show her, was Peter, that neither did he.

But of course he cared very much; and he was so full of wrath against grown-ups, who, as usual, were spoiling everything, that he breathed intentionally quick short breaths at the rate of about five a second. He did this because there is a saying in Neverland that every time you breathe, a grown-up dies; and Peter was killing them off vindictively as fast as possible.

Then he went to Tinker Bell's niche in the wall and said, "You are to get up, Tink, and take Wendy, John, and Michael back to their home."

Tink had been sitting up in bed listening for some time, and she was delighted that Wendy was going.

"Now then," cried Peter, "no fuss, no blubbering; goodbye, Wendy," and he held out his hand cheerily, quite as if they really must go now.

She held out her hand, as there was no indication that he would prefer a thimble.

"You will remember about changing your flannels, Peter?" she said, lingering over him. She was always so particular about their flannels.

That seemed to be everything, and an awkward pause followed. Peter, however, was not the kind that breaks down in front of people. "Are you ready, Tinker Bell?" he called out.

"Aye, aye."

"Then lead the way."

Tink darted up the nearest tree; but no one followed her, for it was this moment that the pirates made their dreadful attack upon the redskins. Above, where all had been so still, the air was rent with shrieks and the clash of steel. Below, there was dead silence. Mouths opened and remained open. Wendy fell on her knees, but her arms were extended towards Peter. All arms were extended towards him, as if suddenly blown in his direction; they were beseeching him mutely not to desert them. As for Peter he seized his sword, and the lust of battle was in his eye. But as sudden as it had begun, the pandemonium above ceased.

Which side had won?

The pirates, listening avidly at the mouths of the trees, heard the question put by every boy, and alas! they also heard Peter's answer.

"If the redskins have won," he said, "they will beat the tom-tom; it is always their

161

sign of victory."

Now Smee had found the tom-tom, and was at that moment sitting on it. "You will never hear the tom-tom again," he muttered, but inaudibly of course, for strict silence had been enjoined. To his amazement, Hook signed to him to beat the tom-tom; and slowly there came to Smee an understanding of the dreadful wickedness of the order. Never, probably, had this simple man admired Hook so much.

Twice Smee beat upon the instrument, and then stopped to listen gleefully.

"The tom-tom," the villains heard Peter cry, "an Indian victory!"

The doomed children answered with a cheer that was music to the black hearts above. They smirked at each other and rubbed their hands. Rapidly and silently Hook gave his orders: one man to each tree, and the others to arrange themselves in a line two yards apart.

The more quickly this horror is disposed of the better. The first to emerge from his tree, sword in hand, was Peter. But Cecco of the foul garlic breath quickly stepped from behind his tree and disarmed him. Then Cecco flung him to Smee, who threw him to Starkey, who flung him to Bill Jukes, who flung him to Noodler, and so he was tossed from one another till he fell at the feet of the black pirate. All of the boys were plucked from their trees in this ruthless manner, and several of them were in the air at a time, like bales of goods flung from hand to hand.

A different treatment was accorded to Wendy, who came last. With ironical politeness Hook raised his hat to her, and, offering her his arm, escorted her to the spot where the others were being gagged. He did it with such an air, that she was too fascinated to cry out.

There were all of them tied to prevent their flying away, doubled up with their knees close to their ears; and for the trussing of them the black pirate had cut a rope into eleven equal pieces. But wait! Should there not be twelve?

Below they numbered the same as the disciples. Hook will live to regret overlooking the twelfth. The black pirate's Judas' trick has flushed them out save one—the baby. Upon hearing the first sounds of warfare from above, he had dived to the bottom of his basket, and there had not heard Peter's announcement of an Indian victory, and so was still deeply burrowed into his bedding when the others went above.

Bertie, when he realized Bongo was missing, did not want to be separated from him, but thought the better of telling the pirates of his whereabouts as they were carried off to the pirate ship.

Let us pretend now to steal on board that dread pirate ship where Peter and Wendy, the lost boys and Bertie are held prisoners. The slightest of winds—no more than a zephyr—wrinkles the placid face of the lagoon. On deck Smee has the watch. His eyes dart this way and that, and his hand, anxious for whatever excuse to draw his sword, falls from time to time upon Johnny Corkscrew hanging from his waist.

Below, in a dimly-lit and stinking cabin, Wendy and the lost boys are guarded by a single pirate. Disarmed, they are of no consequence to Hook. Tied so securely to the after mast that he can hardly wiggle a toe is Peter Pan. And lashed just as firmly to the forward mast is Bertie, who for a reason that we shall reveal forthwith, has attracted the sinister attentions of Hook, emerging now from the stairway from below.

It is a fine night, even for the tropics in Fairyland. Rising over the open sea, the full moon coats the lagoon with a film of silver. Hook pauses to take in the evening's beauty, then stalks over to where Bertie is encircled by strands and strands of Hook's strongest hemp. Since being tied here, Bertie has pondered his separation from the others. Hook's antipathy to Peter he has heard from Pan's own lips, so he did not wonder when he saw Peter tied to the mast. But then, as he and the other children were being marched below, Hook's eye fell on him like a shadow, and he was pulled out from the pack like a sacrificial lamb from a drove of sheep.

Now Hook paced slowly back and forth, observing him from front, back, and sides, pausing at each perspective. Presently, he sat down behind him.

"You have been rolling around in your mind, I suspect," Hook said, "hoping to strike a nine-pin of meaning, my reason for isolating you from the rest. How you come to be on the island I do not know, and before today I know you not. Tomorrow, however, will be your last day." He paused now to enable the dire meaning of his words to sink in.

"Tomorrow you and all the other troublesome boys will walk the plank. I venture the crocodile will be waiting alongside to make a mangled meal of you. The girl we shall keep for mending and the washing up." Again Hook paused.

"What I am about to tell you now, no other mortal soul knows. After tomorrow, it will be of no consequence, since you will not be able to reveal it to anyone." Again he paused, as if contemplating how much of his life he could reveal.

"This coarse pirate who stands before you is not what he appears. I am of titled stock, related to royal blood—*English* royal blood. At about your age," he said to Bertie, "I commenced my schooling at a famous public school. I came away from it with honors—both in the classroom and on the playing field. And most of all, I came away with a passion for good form."

Nightly there came from far within his head a creaking as of rusty portals opening, through which there stalked a stern tap-tap-tap, like a blind fate pursuing him with an inexorable question. "Have you been in good form today?" was the question.

"Fame, fame, that glittering bauble, is mine," he answered. "I am the only man whom Barbecue feared. And Flint the great and terrible himself feared Barbecue."

"Barbecue? Flint? What school, which house?" came the cutting retort.

Hook's most disquieting reflection: was it not bad form to be preoccupied with good form? But the more he tried to put it out of his head, the more it came tap-tap-tapping, hammering at him in the night when he could not sleep. It was a claw within his vitals, sharper than his iron barb, and it raked his conscience up and down until the sweat ran in rivulets down his sallow countenance and streaked his jacket. Oft-times he drew his sleeve across his grimy face, but there was no stopping that damning sweat.

"Was it not bad form to sweat?" came the annihilating question.

Hook was beginning to sweat profusely in Bertie's presence, and now in the twinkling of an eye he knew why. For a moment, he did not think that he could bare himself further to the young boy. Then, consoled by the knowledge that after tomorrow he alone would be once more the keeper of his secret, Hook continued.

"After school I joined the Royal Navy with a commission. My first ship was *H.M.S. Indomitable*. Have you heard of her? No matter. The captain's name was Vere, but it is the Master-at-Arms with whom my narrative is concerned. John Claggert was his name, or Jimmy-Legs, as the men called him, behind his back, of course, because he was the strictest taskmaster to walk the seven seas. I soon found that out. As an officer, I did not come under his jurisdiction, but I saw first hand how he worked.

"Then one day we encountered an English merchant ship and pressed into our service one of her crew, a young seaman named Budd. Your resemblance to him is remarkable, but his hair was fairer and his eyes bluer; only for a moment I thought he had come back again. From the moment he came on board, he made all the crew love him, all save Claggert. He was like a sky god, all golden curls and eyes bluer than the sky. He was simplicity and goodness personified. But he was *too* good—thought ill of no man; yet his life was an example no man could follow.

"Claggert was his opposite. Jimmy-Legs was like the sea at its darkest depths, all blackness of mood and motive. And the sea was in the strange poetry of his speech. Beneath its calm surface he would say, lay monsters preying on their fellow beings. 'The sea's deceitful, Billy Budd,' he warned him, 'tranquil, unperturbed above, and roiling with blood below.'

"Claggert was warning Billy not just about the sea; no, indeed, he was warning him about himself, that devil in a man. But Billy could not accept evil, nor see it in any man. Fearing him and hating him, the crew were all in Claggert's power, except Billy. When their hate of Claggert turned to love of Billy, it stripped Claggert of his power, left him feeling impotent and angry, so he set out to destroy Billy Budd.

"At first it was only petty things, sending his lackeys to mess up his gear so that he could be put upon report. Then he tried to tempt him with gold to lead a bogus mutiny, but Billy would have none of it. You see, he was incapable of doing or even thinking evil in any form. He was an innocent in every way, and this made him a holy fool.

"Finally Jimmy Legs could stand it no longer and played his trump card with the captain, accusing Billy of spreading rebellion among the men and urging them to act with him in mutiny.

"The captain believed not a word of it, for he was as won over by Billy as the rest of us, and had already promoted him to captain of the foretop after his commendable conduct in an action with a French frigate, which we chased, but she eluded us.

"To squelch Claggert he brought Billy before him and told Jimmy-Legs to repeat what he had said to Billy's face. Suddenly Billy, who could think ill of no man, was face to face with utter evil. His whole world came apart. I was there with the captain and another officer. Billy could not speak, despite the captain's urgings for him to defend himself.

"'For God's sake, Budd, deny this evil!'" screamed the captain.

"Billy desired greatly to oblige the captain, who had treated him like his own son, but the more he tried to speak, the more he stammered. His body contorted in dumb gestures, and only gurglings came from his throat. It was horrible to see. Good and evil, face to face, and evil was triumphing. Finally his agitation overcame him, and Billy lashed out with a single blow that caught Claggert in the side of the head and snapped his neck."

During all this narrative, Hook had been seated behind Bertie. Now he rose and went round in front of him where he began to pace up and down as he spoke.

"Moments later, Captain Vere convened a court, comprised of the officers on board, to try Billy Budd for striking and killing his *superior*, John Claggert. Hook spoke the word *superior* with a mouth full of irony. To a man, we all rallied to Budd's cause. The concensus was he struck in self-defense, prevented by his speech impediment from *verbally* defending himself. We were of one mind, but there had been *two* mutinies aboard English ships just previous to this time, and the Captain played devil's advocate, or Claggert's, that is, by making us aware that a verdict of innocent would be unacceptable to King and Country.

"In the end, we gave in. I can remember saying to the captain, 'This is tyranny, not law, to give the victory to the devil himself.'

"And so Billy Budd was judged guilty of the capital crime of killing a superior officer. And the punishment for a capital crime is death. I helped set the noose around Billy Budd's neck. I tell you we very nearly had a mutiny then, but Billy saved the day once more, shouting, 'God bless Captain Vere,' as he stepped out into an eternity of open space."

Hook is silent now, ceasing his pacing. Presently he kneels down on the deck in front of Bertie. Opening his doublet, he reveals a wide leather belt into which is burned the initials BB.

"This is his belt. He gave it to me before we put the noose around his neck. Imagine! When we came ashore, I resigned my commission. Signed on with a merchant ship and jumped ship in the Caribbean. Now I am the blackest pirate that ever sailed the Main! But the man that stands before you now was not always so black, no, indeed.

"You see, I told myself if the law could hang the best of men, then the law was wrong. And if the law was wrong, the entire system was corrupt. So I set myself against it, totally and without compromise. I have no regrets, and tomorrow my final goal will be accomplished—the utter destruction of Peter Pan."

Hook laughs aloud, then stands again and begins pacing up and down in front of

Bertie, never looking at him, and speaking as if defending himself to the world.

"Sometimes I ask myself why a mere little boy should so arouse my wrath—I, the greatest pirate of all! It is his unbearable cockiness. He *is* good, I grant that, but to do something well and then *crow* about it, that is the worst of bad form. You know what bad form is, eh, eh, eh?

"To do well without humility nullifies the doing. Now take the example of Billy Budd. He did well with *proper* humility. Good form always. That is why we all looked up to him. He set an example. But what an example! No mortal could emulate his life. And he forgave us at the end, forgave us our law which said we must kill him, and forgave us his death. But I tell you, son, *I* began to die when I saw Billy dance at the end of that rope.

"What is left for me now? With Pan gone tomorrow the last canker in my craw is cut out. Peace at last? No, the British Navy will one day find me. I cannot win. And it would be the poorest of form to finally triumph. History only remembers the triumph of good over evil. All else is swept under the carpet.

"Shall I tell you how I foresee my end?" Hook pauses and looks off to the moonlit horizon, as if gazing into the future.

"Classics was my field at school. In the Egyptian religion just as in our Christian religion, there is a judgment of the dead, presided over by Osiris, their great god. The goddess of truth and justice stands ready to weigh the heart of the deceased. Those that fail the judgment have their hearts devoured by a crocodile-headed god.

"You have seen the crocodile that stalks this island. He pursues me relentlessly, thanks to Peter Pan. He has tasted my arm and now craves my very heart. Only the ticking of a clock he swallowed alerts me to his presence. One day time will run out for me, and then I shall be judged, and found—I hope—in good form at the last."

Now Hook kneels again before the bound Bertie. For a long moment he studies his face carefully.

"Perhaps it's only a trick of the moonlight, but it's uncanny.... Your resemblance to Billy Budd. That same open, guileless face. Uncanny. All right, let me save your life this time.

"The ship can accommodate a cabin boy. Your duties would not be overwhelming, and your treatment not harsh—I'll see to that. If anyone harms you, I'll cut them with my iron claw. What say you now?"

Hook peers intently into Bertie's face.

"I would like to stay," said Bertie, "but I have a purpose."

"Eh? What purpose?" replied Hook warily.

"It's a long story," said Bertie, "I have to be free to get back in space/time to save my mother and father from a flying bomb."

"Flying bomb?! From where?"

"Out of the sky. From Germany. It killed my mother and father once in the past, and now I have to get back there somehow. Please let me go."

"Let you go, to blab my story all over England? Never."

"Please. Your time is way before mine. People won't remember your story. They won't care anymore."

"Not remember, not care about my story! You're saying I'll be forgotten by then?! They'll always remember Hook the great pirate!"

"Please, you don't understand. You're making another mistake. You had a *choice* for good or evil when you met Billy Budd. You could have stayed in the navy and worked to change the system. You didn't have to try to be as bad as you could be. You became just like Jimmy-Legs, and his evil goes on."

"Me?" said Hook, incredulously, "like Claggert. I hate him."

"Your hate binds you to him," said Bertie. "Tom hated Grimes, and he had to go to him and reward evil with good. One day Mrs. Bedonebyasyoudid will be coming for you, and you will have to find Jimmy-Legs and do something for him."

"I'll do something for him if I ever meet him again," said Hook, brandishing his iron claw. "I'd lead him a merry dance on the coals of hell. Say, who are these people, Tom, and Mrs. Didby?"

"They're in *The Water-Babies*. Didn't your mother ever read it to you? It's a wonderful story."

"What's a story got to do with me then?" asked Hook.

"That's just it. I don't belong in this story, and I have to get out so I can go on with my purpose."

"So Hook's become famous enough to have his own story, eh?" he said, smiling. "Well, then, how does it turn out for me?"

"No, it's not your story. It's called *Peter Pan*."

Hook spat out the words "Peter Pan."

"Yes, you see your story comes way before my life. It's already been finalized. There's even a statue of Peter in Kensington. I couldn't hurt your reputation as a great pirate by talking about you, because your fame is made."

"Don't try to butter me up, laddie," said Hook furiously. "If they've erected a Peter in Kensington, then I've one more score to settle with the British nation, and one more with him. As for you who bears the likeness of Billy Budd, either you agree to be my cabin boy or you walk the plank with the rest!"

"No!" said Bertie.

Hook laid his iron claw against Bertie's throat, and thrust his face into his. "Damn you, are you going to die again at my hand?" Then he turned and stalked away.

For what seemed a very long time Bertie watched the moon rise higher and higher. His arms and legs had become numb from the tightness of the rope, and the moon passed over his head and out of his vision when he at last fell into a fitful sleep.

Was he dreaming, or did he hear Mrs. Doasyouwouldbedoneby.

"Every moment is an enlightenment," came her words. Were they crossing the

167

sea between two stories, or were they inside his head?

"Those who would be clean, will be clean," came her voice again. Bertie thought now of Tom, and how he had struggled to free himself from Grime's grasp in order to bathe in the cool limestone pool. Now Bertie tried to wiggle against the weight of the ropes, but only his feet and head could move so securely had the pirates tied him.

Then Bertie thought of something Mrs. B. had said: the nature of darkness is to want to control the will of others. A dark star controls its light, came the next thought. And Bertie began to wonder how many dark stars he was looking at that gave no light. Suddenly it dawned on him that tied to the mast as he was, he was like a dark star. He had no freedom, no choices. No light could escape the star; Bertie could not escape. His spirits began to sag, and his upper lip to quiver when a new star rose over the horizon.

Lo, and behold, it was Tom's dog, set up in the place of the old dog star. Bertie felt a new warmth at seeing his old friend. It was as if the rays of light from the star were penetrating Bertie's darkness of spirit, bringing new comfort. Presently there came to him a vision in the sky of Tom and Ellie, grown-up and happy, standing with their arms around each other and beaming down love on him. Then he fell asleep.

Sometime much later he felt a gentle tapping on his shoulder. He could not turn his head, and it came again, more insistent. Then he felt a wet, fuzzy head snuggle against his cheek. It was Bongo!

When the pirates had taken the boys prisoner, he had hidden unnoticed under the bed. Swimming under cover of darkness, he had just now boarded the ship. Bertie was so glad to see his friend that he forgot for a time that tomorrow he would walk the plank. Then a plan began to form in his mind.

"Now listen carefully," he said, and Bongo cocked his head attentively and folded his hands patiently in his lap as Bertie spoke long and quietly to him.

That morning the sun dawned blood-red. A few early-rising pirates emerged on deck, yawning and stretching. Presently Hook's voice was heard from below, angrily issuing orders. One by one the boys were brought on deck, a pirate guarding each one.

"Hallo," said Tootles, "there's Bertie. Have you been there all night?"

"Poor chap, he can't even move his head," said John.

"Don't worry," called Nibs, "Peter is coming to save us."

Just at that moment, Hook emerged from below, laughing mockingly. "Aye, he'll be along any moment now. Smee, Starkey, come with me. We'll cut the cockadoodle's ropes, and never you mind if you prick him; it'll sweeten the water for the sharks."

The boys are herded past Bertie to the most forward part of the deck, where Cecco and Bill Jukes are rigging the plank. They test its spring, then pronounce it ready.

Starkey and Smee, followed by Hook, drag Peter along, a hand on each shoulder. Hook stops in front of Bertie, as the others go on.

"Is it to be my cabin boy?" he asks.

"I already told you I have a purpose," says Bertie.

"Then we are at cross-purposes and you must walk the plank," says Hook, slashing at Bertie's ropes.

When the boys have been herded onto the forward poop deck, all the pirates withdraw to form a wide line behind them, stretching from port to starboard rails in order to catch them should they break and run. All, that is, save Smee, who has unsheathed Johnny Corkscrew, and will use it as an inducement to moving the boys along the plank. At a signal from Hook, the other pirates draw their weapons.

"Now do it nice and neat lads," Hook calls out. "Don't let the old school down. Step smartly out on the plank one by one. Don't make us have to cut you up on deck."

"Man'o'war on the horizon!" comes the cry. It is Bertie's voice, and he calls again. "Man'o'warrr!"

The pirates turn and look to a man.

"Eh, I see nothing," cries Hook. Just then a puff of smoke and a fiery explosion shatters the deck beside him. Hook and two of his cohorts are catapulted into the air. Before they have a chance to recover, another flash falls within their line and two more pirates fly through the air. Then flares, eruptions, and detonations with pinpoint accuracy decimate their ranks.

"Abandon ship!" calls Bertie, as those pirates who still can leap over the railing and into the sea, while the boys stand watching their antics dumbfounded, jaws agape. Smee drops Johnny Corkscrew and follows his mates overboard.

As the smoke clears, Wendy is seen running barefoot along the deck into the waiting arms of Peter Pan, who now holds Johnny Corkscrew. Hook alone is left of the pirate crew.

"I give the orders on this ship," he said, reprimandingly to Bertie.

"What ship has saved us?" cry the boys, squinting into the sun rising over the ocean.

"H.M.S. Bongo," calls Bertie, pointing aloft. From the crow's nest, Bongo is seen waving.

"Three cheers for Bongo," cries Peter. "Hip, hip, huraay! Hip, hip, huray! Hip, hip, huray!"

Then as Bertie watches horrified, Bongo raises another coconut overhead.

"No, no, Bongo," cries Bertie. Bertie cringes and falls to the ground. The coconut shatters harmlessly on the deck.

"I'll get you for that," laughs Bertie, and Bongo starts down from the crow's nest.

"What trick is this?" says Hook.

"Coconuts filled with gunpowder and a fuse through the eyes," says Bertie.

"Lost my ship to a monkeyshine!" exclaims Hook.

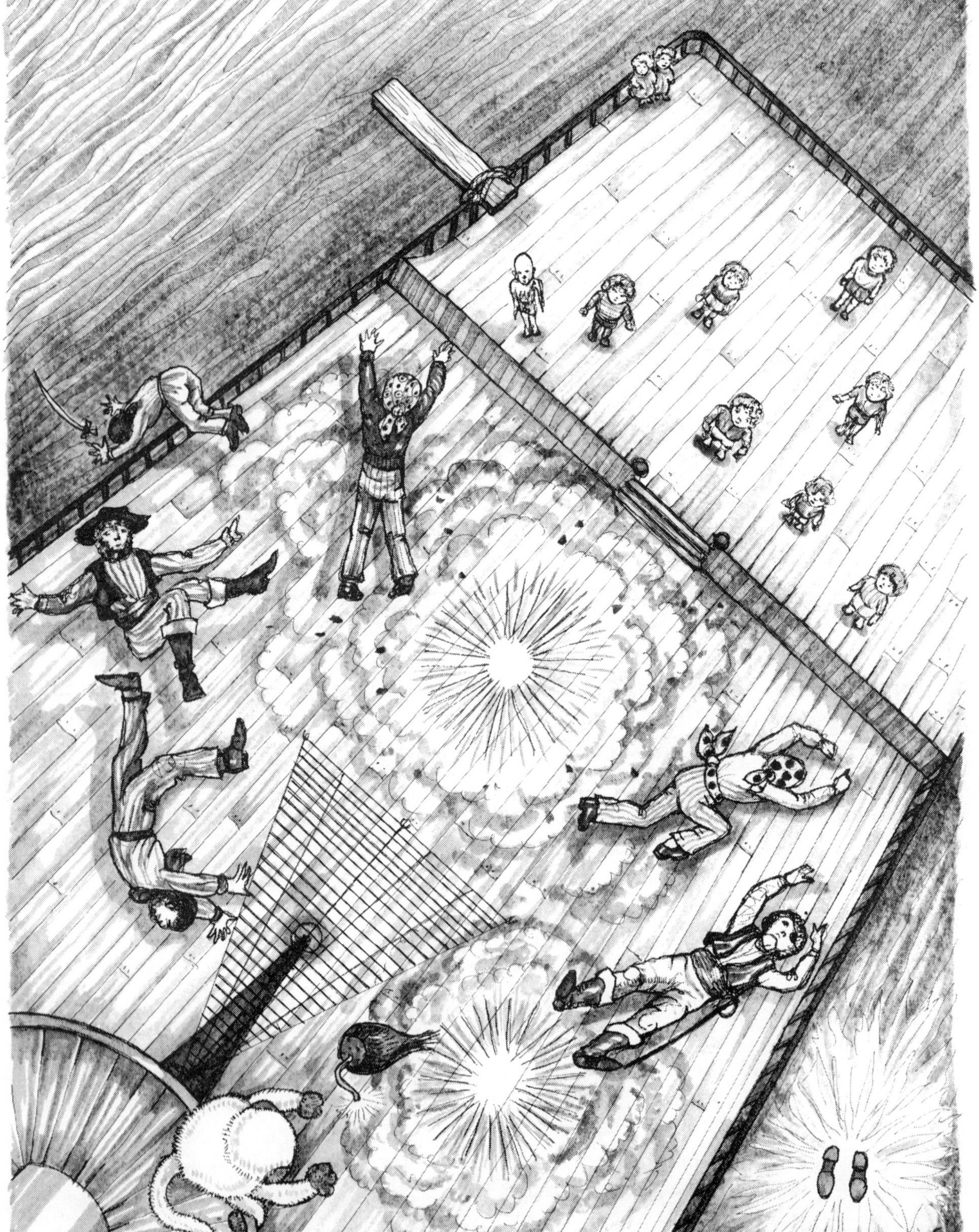

The boys gather around the mast, and as Bongo comes down he leaps the last few feet into their waiting arms. Peter remains where he had been, holding Wendy. Tootles places Bongo on his shoulders and they parade him around the deck, chanting "for he's a jolly good fellow."

Peter has cast a cold eye on these proceedings. Loving the limelight as he does, he views Bongo's triumph as a challenge to his authority.

"Tie up Hook!" cries Bertie to the boys. They hesitate, Peter eyeing them coldly.

"I give the orders here," he barks.

"He has to be turned over to the authorities," says Bertie. "He's wanted for crimes against the Crown."

"On this island, *I* am the authority!" shouts Peter. "Hook or me this time," he cries and launches himself at Hook.

At first Hook seems warm to his work, glad for the chance to have at Peter. Victory, the destruction of Peter Pan, may yet be snatched from the jaws of defeat. But Peter proves more than what Hook bargained for. Remarkably adroit, Peter Pan has long craved this encounter. He dances about Hook like a will-o-the-wisp, flicking his sword into the wrist, arm, and sides of the furious Hook. The boys form a circle, shouting their encouragement. Wendy tries not to look, but cannot should she be needed to intercede if Peter is wounded. Flailing his hook and a sword in his one good hand in wide arcs, Hook tries to catch Peter with a lucky stroke, but it is no good. He darts in under or after each stroke like a stinging bee, until Hook is slowly driven to the rail. In one last attempt to gain the advantage, Hook springs upon the rail, but Peter leaps easily up. Hook tries to turn but teeters uncertainly, his back to Peter. Peter unleashes a savage kick, and the tottering Hook, his balance lost, begins the long fall to the sea below.

"Bad formmm," he cries jeeringly to Peter.

Now in his mind he *had* the victory after all. He was no longer the pirate; no longer *there*. Back upon the playing fields of his old school, playing a game atop a famous wall, into whose stones the great and famous had carved their names as boys playing the game, with his shoes and sox right, his tie and waistcoat perfectly proper, he had been *kicked* from the wall by an upstart boy who personified bad form.

And then, in the last instant of his life, he saw it. The crocodile, jaws agape! He fell in and disappeared down a black hole.

The next morning, when the damage to the ship had been assessed, Bongo's bombadillos had done no more than blacken the deck where they struck. Wendy called for a grand scrubbing of decks and cabins alike, and Peter grudgingly granted the clean-up. With an early start, by mid-day they had finished.

Reconnoitering the ship, they found no real treasure aboard, except for some fancy rings and bracelets, which Peter gave to Wendy, since he did not deem it manly to wear them himself.

There were, however, some rather extraordinary garments closeted in the cabins: capes, elegant hats, pantaloons, tunics, frilled shirts and such, probably from victims who had been stripped to their drawers before walking the plank. All children love to dress up in adult clothes, so the afternoon was given over to choosing costumes. The sleeves could be turned up, but the pants were too long to wear without tripping up. Given the task of cutting the trousers at the knees, Wendy was at the rail of the ship shaking out a dusty pair of breeches when a curious event took place. Tinker Bell's thimble, given her by Peter, shook loose from Wendy's finger and dropped over the side.

"Oh, noooo!" she cried in dismay, watching its fall as if somehow she might retrieve it if she marked the spot. But before it could enter the water, a golden carp surfaced, swallowed the thimble, and disappeared beneath the waves. Wendy was so heartsick she could say nothing to Peter. And a kind of enchantment, a betrothal to Neverland, had been broken.

After supper, sitting around in their new guises they held a pow-wow. It was apparent that Peter itched to return to the island. If he no longer had antagonists, he still had the worship of the Indians, to whom he was a kind of demi-god for rescuing Tiger Lily. Scorning their flamboyant grown-up costumes, he still clothed himself sparsely in leaves alone.

"I'm for blowing up the ship, and returning to the island," said Peter.

"Jolly good!" cried the boys.

Now it must be observed that boys everywhere love a show of fireworks, no matter the costs, and the boys of Neverland were no different.

"We'll blow her sky high!" cried Tootles.

"Dibbies on lighting the fuse!" exclaimed Nibs.

"Oh, Peter," pleaded John, "couldn't we blow her piecemeal—a part each day."

"Capital idea!" seconded Slightly. "A deck a day, say you Peter?"

In the excitement of possibly walking the plank, Wendy had chosen to forget for a time that she was on her way out of Neverland when the pirates attacked. Now the prospect of returning with Peter to the snug home underground seemed positively claustrophobic. Stirring within her were the first desires for motherhood, which is not the same as mothering. Peter wanted mothering, because he was determined to remain a boy forever, whereas motherhood meant for Wendy one day marrying and mothering children of her own. She could not imagine mothering Peter forever, and since he refused to grow up and become a proper father she could include him in her plans no longer.

"Stop!" said Wendy. They all turned and looked at her.

"All these plans for going back to the underground house, when you forget, John and Michael, that we *were* on our way home.. After all, a house is not a home."

"Oh, but why go now, Wendy," pleaded John. "With the pirates gone, there aren't any adults around now to spoil the fun."

"You're starting to think like Peter, and that's a very good sign that we need to leave here. Now take off those costumes and get ready."

Here he had just defeated Hook in a great duel, and she was still thinking of domesticating him. For Peter, it was the last straw.

But Bongo looked so disheartened at the departure of his mother, balling his paws into his eyes, that Wendy nearly relented for his sake, but then had a bright idea.

"Bertie and Bongo, would you like to come with us. After all we are going to London."

Bertie reflected for a moment, while Bongo indicated that he very much wanted to keep Wendy for a mother, but then he said reluctantly, "Well, the *place* is right, but the time is wrong. I have to go back to 1945, and *your* time is too early."

"Peter, can we go with Wendy," spoke up Nibs.

"Dear ones," she said, "if you will all come with me, I feel almost certain I can get my mother and father to adopt you."

"But won't they think us rather a handful?" Nibs asked.

"Oh, no," said Wendy, rapidly thinking it out, "it will only mean having a few beds in the drawing-room."

"Peter, can we go?" they all cried imploringly.

"Peter, you come too," said Wendy.

"No."

"Yes, Peter."

"No, I will not go with you, Wendy." Then he looked at Bertie. "Since you don't want to go with her either, I'll make you captain of the Jolly Roger, and Bongo can be your first mate." Bongo saluted smartly. "The rest of this lot can fly back with Tinker Bell, for all I care."

"Pctcr, I want you to come."

"And what would I do once I was there?"

"Well, go to school with the other boys, I suppose."

"That's what I thought. Then I'd run away again to Kensington Park, only there are no Indians and pirates there like Neverland, so where's the adventure? And every day *there*, I'd be getting a day older, bumming shillings to eat with, and sleeping on the park benches under dirty newspapers. Do you think they'd erect a statue to me then?!"

"Well, I must say you don't paint a very pretty picture," said Wendy. "As I see it, you're totally irresponsible."

"Irresponsible?" said Peter. "You think that's why I don't want to grow up. Who do you think runs this island. When I'm not here, it's as if Neverland sleeps. The whole show is responsible to me."

"But why are you afraid of growing up?" Bertie asked.

Peter seemed to grow pale for a moment, then he sat down on the deck of the ship. "All right, I'll tell you," he said. "I don't know exactly, but it has something to do with the clock in the crocodile. When Hook was alive, I was not afraid. The croc was after him not me. I gave it Hook's arm, so that it would go after him instead of me. What happens to people when they grow up, Wendy?"

She gave him a long, final look as she answered his question. "Well, as I have said, they become responsible. They marry, and have children of their own, not just pretend children. And they go to real schools, and learn real things, and live in the real world."

"You forgot one thing about growing up," said Peter.

"Oh?" said Wendy, and all the children waited for his answer.

"They become old. And when they become old, they cannot fly anymore. If I couldn't fly, I wouldn't be *free*!"

"I see," said Bertie, "and in a way I envy you. Another time I might have stayed here with you—a new adventure every day, but I know I have a purpose, and you can help me if you will."

"How can I help you?" Pan asked.

"This ship can help. Ratty gave me a boat that got me out of *The Wind in the Willows*. This ship could carry me and Bongo to another place."

"It's rather a lot to ask," Peter replied, "but if you think it will help, I'll not deny it to you. We can help you rig it, and with the wheel tied down while you slept, the trade winds would blow you towards England. These are all Westerlies in this latitude."

"Three cheers for Peter, then!" cried Bertie.

"Hip-hip-hooray! Hip-hip-hooray! Hip-hip-hooray!" and they all joined in, save Wendy who now realized that she *had* to go. Before it had been a kind of game that she played with Peter, *pretending* to leave. Now it was too late.

As if in answer to Peter's promise to Bertie, the coconut palms swayed to a fresh new breeze. In half an hour of smart work, the *Roger* was readied, and the skull and bones lowered and the British flag raised.

Then it was time.

"No tricks with them, Tinker Bell. Fly straight to London."

The boys all stood on the ship's railing, anxious to get a flying start. Peter went to Wendy and offered to shake her hand. She hesitated.

"Peter, you know that thimble you gave me from Tinkerbell? Well, I'm very much ashamed to say it fell overboard, and I'm wondering if you'd give me one of *your* thimbles as a goodbye present."

Standing on tip-toe, Peter pressed his lips against Wendy's. The boy's smirked and looked away. Kissing girls was not something *boys* did, and they were embarrassed for him.

Wendy stepped onto the railing with the others. "I shall remember that even when I am old and grey," she said. Then Tinker Bell shook fairy dust on each of them, and they rose into the air, waving at Peter, Bongo, and Bertie on the ship's deck until they were too small to be seem.

Pan helped Bertie weigh anchor, and the ship began to sail.

"Thanks awfully, old man," said Bertie, shaking hands.

"Don't mention it," said Peter, "and don't call me 'old man.'"

"Sorry," said Bertie, "I forgot."

Then Bongo offered his hand and Peter shook it.

"Wouldn't like to stick around as my mascot, would you?" Peter asked Bongo, who emphatically shook his head 'no'.

"Well, then," said Peter, climbing onto the railing, "until Neverland comes round again," and he dove into the water. Bertie and Bongo watched him until he emerged on the beach and disappeared into the jungle.

Gradually the island sunk on the horizon like a setting pancake. The stars came out. Tom and the dog were overhead to guide them—home or where?

GLOSSARY
BOOK THREE

adhered: followed
adroit: clever and quick
aghast: amazed, frightened and shocked
annihilation: destroyed to the point where nothing remains
antagonists: those who like to cause one problems
antipathy: dislike
ardently: a warm and glowing feeling
betrayed: a secret revealed
betrothed: fiancee, one promised to marry
blackavised: dark looking
bosun: short for boat-swain which means boat-boy; his duty's to keep the boat from falling apart
breach: gap-a break, large or small
brooding: worrying
cadaverous: dead looking
cast: family-essence-what one is made of
commenced: began
commendable: worthy of praise
conceited: smug
consoled: comforted
contemplating: thinking deeply
contemptuously: sneering
contrivance: invention
countenance: looks and appearance
dative: indirect object of a verb, if someone sends you an invitation,   <u>you</u> have made a dative, if you do not <u>object</u>
debonair: graceful and charming
decimate: making a thing smaller by destroying it bit by bit
deem: thinking that things are this way or that
demigod: half human, half god
diffidently: shyly and timidly
discerned: noticed
domesticating: taming
ecstatically: happily times a thousand
eerie: strange and a little frightening
eluding: sneaking by
emphatically: absolutely positively!
enjoined: ordered
exquisite: elegantly beautiful perfection
fastidious: carefully neat
fey: of another world, perhaps of fairyland
flamboyant: colorful
foreboding: warning of unhappy things
gnashed: rubbing one's teeth together, which is sometimes charming
guises: the clothing of other selves, costumes and masks
harpsichord: a piano's great grandmother
impede: something in one's way is an impediment
imploringly: begging
impotent: powerless
incredulous: the way one feels when things are almost completely impossible to believe
indignantly: anger when things seem unfair

inducement: persuasion
inexorable: something that will not stop, however much we may wish that it would
inflections: the way one says things
inscrutable: almost impossible to understand
insurrection: rising up against rules and authority
intercede: coming between things, perhaps to help
interposed: interrupting a conversation with one's own ideas or adding one's own two cents
irony: deeper meaning
jot: a tiny little bit, somewhat more than a smithereen
jurisdiction: the area in which one is allowed to make all the rules
limelight: center of attention
lubber: (lover) land lubber-pirate term for those who do not know the sea
malice: unfriendliness
melodious: lovely, like a song
mournful: sad
nullified: something which is turned into nothing
overtures: making friendly motions, usually to a stranger
Pandemonium: wild uproar
perambulators: prams-baby carriages
petulantly: cranky, and in a rather bad mood
piecemeal: piece by piece and bit by bit
pluperfect: When a thing took a little time to pass in the past, rather than just things past
port: left side of the ship, when facing forward, which is the bow
prattle: chatter
profusely: in large amounts
prowess: skill and talent
psyche: inner mind and soul
rakish: handsome
rampageous: wild rough and tumble, a love of knocking things about
reconnoitering: looking things over
recriminations: blaming and accusing
reflected: thinking while remembering
rent: suddenly torn and broken
reproved: scolded
repute: reputation or public respect
retort: a rather sharp reply
ridicule: accusations and blame
sallow: an unhealthy grayish greenish yellowish color in the skin
sarcastically: saying something one doesn't really mean, usually in a rather rude way, to imply what one really
    does mean
scandalized: shocked to think of such a thing:
scrutinized: looked very closely
stalked: walked angrily
starboard: right side of the ship, when not facing backward, which is the stern
steal: sneak
sublime: deep and true
tentatively: hesitantly
trump card: a secret which if revealed at the right moment may allow one to win the game
uncanny: strange and weird
vindictively: reacting angrily from within a grudge
weigh anchor: pull up the anchor and set sail
will-o-the-wisp: like a wind or a ghost, appearing for a moment then disappearing again
zephyr: peaceful and gentle wind
betrothed: fiancee

BOOK FOUR

To the supple woman
Marry the upright man.

Wood

Came they unto us in the year of our Lord Eleven hundred and sixty-three, as prophesied by Merlin, for our deliverance from the pillage of the Norman lords who held us in thrall nigh an hundred years, since the coming of William named "Conqueror" from Normandy in France, where he slew Harold, son of Edward called "Confessor," whom death attended like a bride in the first week of Januarius, Ten hundred and sixty-six, in the form of a comet trailing a great white train down the altar of eternity. Crowned then that week Harold, King of England, but the long-haired death star returned on Greater Litany eve, twenty-four April, Ten hundred and sixty-six, and shone both night and day, boding doom for Harold and all England.

At Hastings, Kent, fourteen October, Ten hundred and sixty-six, fell Harold, an arrow in his eye. Quartered was he then by Norman knights and his entrails scattered over the battlefield.

Crowned then on the day of the birth of our Lord Jesus Christ, Ten hundred and sixty-six, William the First, King of England.

Crowned then, in the year of our Lord Eleven hundred and fifty-four, Henry the Second, a just and strong man, grandson of William the Conqueror, and castles of the Norman barons he tumbled down, stone by stone, and law and order brought he to our England. But in Nottinghamshire many lords heeded not the King's will that all men be treated fairly, and these barons turned from their lands, which their fathers' fathers had tilled before the coming of the Normans, many a brave and stalwart man, who gathered then in the Shirewood, stretching thirty miles northward from Nottingham and the Trent river.

Such an one was Robin o' the Wood, son of King Henry's Forester, murdered by Norman hand.

It was nighttime and Bertie and Bongo slept while the pirate ship sailed on as if guided by an unseen hand. In those days the Trent River was broader and deeper than now, flooded from the sea as far inland as where it converged with the aptly named Idle River, lazily making its way northward from Sherwood Forest, which was the wood of that shire, Nottingham, and hence called Shirewood.

A big bang and Bertie and Bongo tumbled from their bunk. The ship had run aground on the riverbank. Bertie rubbed his eyes sleepily and looked at Bongo.

"Where have you landed me now?!" he said.

Overhead the full moon cast its phosphorescence onto a leafy canopy that shimmered and twinkled like a thousand stars whenever a gentle zephyr shook the leaves. Like two spellbound dreamers Bertie and Bongo sat and watched the panorama of moonlight and shadow.

"I'm too excited to sleep now," said Bertie. "I want to find out where we are and begin our next adventure. How about you, Bongie?"

Bongo nodded. Overhanging branches on the bank with which the ship had collided provided a gradually descending stairway from deck to ground. Once there, exploring in the night was not the good idea Bertie imagined, for the leafy canopy shut out the moon's light. So that they might avoid walking into something they could not see, or stumbling, Bertie suggested that they go on all fours. And so they proceeded.

"I wonder if this is what Mole felt like underground?" asked Bertie, his nose almost to the ground.

Then he felt Bongo's arms around his neck and knew Bongo felt safer being carried by him. For a long time they proceeded thusly, stopping whenever Bongo's grip tightened too strongly, threatening to cut off Bertie's breathing.

"Not so tight," he would say, and Bongo would relax his grip for a time.

Presently a faint hum became audible through the trees, vibrating like the drone of distant bees. At first Bertie thought it to be a trick of the breeze wafting through the leaves, but then a kind of chant echoed through the wood as if the trees themselves were alive and singing.

"Can it be the Panpipe?" Bertie asked, but since he could not see Bongo's head he did not know his answer.

Following the music like a siren song, Bertie eventually came to a clearing in the forest where a limestone spring bubbled from a deep cave, the interior of which was illuminated by the setting moon, which cast its rays upon a crystalline rock, before which sat a woman spinning an ancient wheel, while around her in groups of three's, stood nine maidens whose singing had lured Bertie to the cave.

The song they sang was more chant than song, rising and falling with strange patterns of scale and harmony, sometimes melodious and silvery as bells, sometimes dark as fear.

Bertie had crept to where he could peer down into the cave without being seen, the moon at his back.

"Pinch me, Bongo," he whispered, for he could only imagine that he was in a dream, but the pain of Bongo's pinch reassured him that he was indeed awake.

Then amidst the chanting that had seemed to him to be in an unknown tongue, three recognizable words stood out, "Empty and fill, fill and empty." Then they were gone in the night like bats flying from the cave, replaced by more words in another tongue, and so the cycle repeated itself, returning eventually to the three English words, "Empty and fill."

Where the shaft of moonlight fell upon the crystalline rock, a second shaft seemed to be drawn from it by the Spinner through the eye of the spindle on her wheel and around its bobbin. The Spinner wore a gold thimble which she used to carefully guide the crystalline thread, taking care not to touch it except with the golden thimble. Although the Spinner sat in full moonlight, half of her body from head to toe appeared to be in shadow.

Without meaning to, Bongo adjusted his position on Bertie's back, and his moving shadow on the rock wall caused nine pairs of eyes to turn upon him. The singing

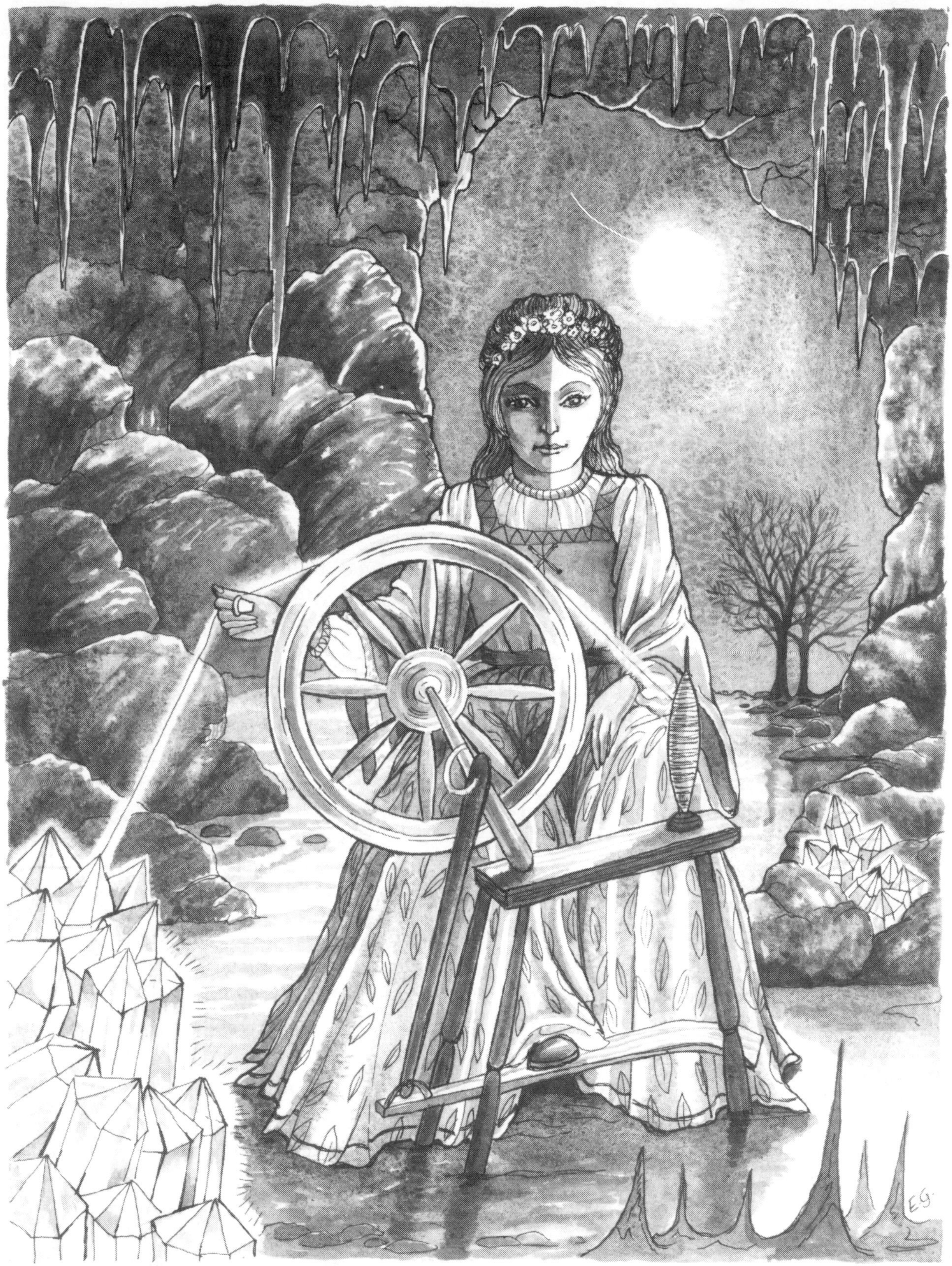

and the spinning ceased.

Bertie was too scared to move. Without turning to look at him, the Spinner said, "Bring him to me." Immediately the singers leaped agilely to catch him. Bertie's brain said, "Move!" but his body remained motionless as if caught in a spell.

When he was carried to the Spinner, Bongo still clinging to his neck, and stretched on the cave floor before her, he felt a strange mix of emotion: fear and love, alternating like an electric current through his body. The beautiful woman now facing him was blonde and fair on one side, and dark on the other with a line marking the division running from the middle of her forehead down the bridge of her nose over her chin and down her neck, disappearing into her garment. She seemed at once old and young, and then Bertie realized that being in her presence was exactly like having present at once Mrs. Doasyouwouldbedoneby and her sister Mrs. Bedonebyasyoudid. One inspired fear, the other love.

Bertie trembled as he lay on the floor, and Bongo shook just as violently, but she held up one finger as if to say, "Fear not," and said, "What is written inside the thimble?"

And he answered, "Marian."

"My name is Marian," she said.

Then as he lay there before her, her eyes moved to the Billy Budd belt which he wore. She indicated it to the other women, and a collective sigh of awe breathed from them. The golden-headed eel or serpent devoured its silver tail.

"Where do you come from?" she asked him.

"From the future," he replied, not yet knowing that he was in the past.

"This is the year of our Lord Eleven hundred and sixty-three," she replied to his not yet spoken question.

"Why am I here?" he asked.

"To help us in our purpose, which shall help you in your purpose," came Marian's reply.

Bertie nodded, knowing somehow as he had known before in the presence of Pan that all would be well.

"Where am I?" he asked.

"In the Shirewood of Nottingham."

"Robin Hood," he said, and smiled. "Robin Hood," he repeated softly to Bongo, who nodded his head enthusiastically. "Is Robin Hood the leader of the merry band?" he asked Marian.

Marian laughed and replied, "Robin couldn't lead them to water," and her nine maidens laughed with her.

Bertie looked puzzled and disappointed, so Marian said, "If you want to meet Robin, you can tomorrow. We will take you there. Now we must leave here, the moon is setting and our night's work is done."

Bertie awoke in a simple hut where he had slept the few hours left of the night before dawn. Outside a voice was calling, "All fires out! All fires out!" The call was repeated by another voice in the distance, and carried on by yet another more distant caller.

Rubbing his eyes sleepily, he looked for Bongo, who was still asleep, his hands cupped together under the side of his head to make a tiny pillow.

"Wake up, dummy!" Bertie said, poking Bongo. "We're in Shirewood forest with Robin Hood, and this is not the time to sleep away our adventure."

Bongo sat up and rubbed his eyes sleepily.

They were in a kind of low-lying chamber, hollowed out under the roots of a giant oak. Straw strewn on the ground comprised their bed. Bertie could stand just barely, but a full-grown adult would have to go on all fours. A deerskin flap separated the chamber from an outer room, from where Marian's head now appeared.

"You are awake, and well it is, for there is much to be done this day," said Marian. "There are some berries, bread, and birch bark tea for breakfast."

They followed her into the outer chamber, which was lit by a single candle in a twisting root of the tree.

"Can we meet Robin Hood?" asked Bertie.

"All in good time," replied Marian, fixing the breakfast. "You must drink your tea while still hot; for I cannot light the fire again until nightfall. The Shire-reeve's men watch all day for signs of smoke in order to learn our position and attack us."

"Is he the same as the Sheriff of Nottingham?" inquired Bertie.

"He is Nottingham's man to be sure," replied Marian. "The reeve is the administrator and enforcer for the county of Nottingham, so he is called the Shire-reeve, but he is beholden to the Earl of Nottingham, the Norman ruler beholden only unto the King."

"And why does the Earl want to have the Shire-reeve attack us?"

"Because," said Marian handing him a piece of bread on which she had spread some red berries, "we are fugitives from his justice. Many of us are free-born men and women whose lands were taken from us, or our fathers, or our fathers' fathers, by the Norman invaders from France. Others of us are serfs, indentured to the land, who could not pay the Earl's tallage tax. Still other free-born men among us could not pay the geld, a tax on one's own land levied by the King."

"My father always said the tax man was the worst," Bertie declared. "But aren't there any real outlaws among you?"

Marian laughed. "Of course, if you don't let the Shire-reeve's men take you, and kill them instead of being killed yourself, then you are an outlaw."

"What about Robin Hood?" inquired Bertie.

"Oh, Robin is always wanted for killing the King's deer," laughed Marian.

"Why did you laugh at him last night when I asked you if he was the leader of the merry band?"

"I don't know why you call us a merry band," replied Marian. "Any day may well be our last. It is only the size of the forest that has enabled us to survive so far."

"But can't Robin defend you? Isn't he the best shot with a bow and arrow in all of England?"

Marian laughed again. "Well, who knows? The bow and arrow are used for killing, either the game to provide food, or to kill an enemy that would kill us, such as the Shire-reeve's men. The King and the Norman lords don't invite us Saxons out to find out who is the best shot in England, and even if they did, none of us would be foolish enough to come."

"Well, that's how they did it in the movies," responded Bertie.

"What nonsense!" said Marian. "What are the movies?"

"Pictures that move and tell a story," Bertie said. "I saw the movie, and you became married to Robin."

Marian laughed heartily. "Well, I can tell you this about Robin. He is no leader of men. He whiles away his time in hunting, fighting, or berry-picking. Set a task for him and he disappears. He is uncomfortable in the company of women, preferring to be alone, or in rough-and-tumble play with his fellows. My marrying Robin, who is at least ten years my junior, is the second most unlikely possibility in all the world, the first being my marrying Thomas Becket, the Archbishop of Canterbury, who despite his previous worldly ways has forsaken the flesh since being made Archbishop by King Henry."

"Well, I don't see how this story can have a happy ending if you don't marry Robin," said Bertie.

"Never you mind about Robin and me!" Marian retorted.

Bongo gave Bertie a straight look, as if to say, "Keep your big mouth shut," and so he washed down the last of his bread with the birch-bark tea and said no more.

Because Marian had much to do, Bertie and Bongo were sent out to play, and soon they were strolling among the great oaks as men came swinging down from the branches or popping up from among the roots, where they had spent the night sleeping.

Hand-in-hand Bertie and Bongo made quite a sensation, for no one in England had ever seen a monkey before. Soon a small crowd of men was following them at a discreet distance. Arguments ensued among the men, which Bertie could hear barely at first, but as they became louder he was aware that they might soon be in trouble.

"They're calling you a devil, Bongo, because of your tail," he said.

The group had grown in size. It seemed that every tree they passed had its own denizens in its branches and roots. As the men drew closer, Bertie commanded Bongo, "Up on my back!" and started running with him as fast as his legs could carry him. But he was no match for the men, and soon lay struggling on the ground, while one man held a kicking Bongo by the tail.

"Look for the Devil's mark!" cried one man, as they pulled Bertie's shirt over his head. Then a collective cry of amazement went up from them as they saw the belt around his waist.

"Billy Budd's belt!" said one.

"To be sure," agreed another.

"What's it doing on him?" asked another.

"Here! Where'd you get that?" demanded another.

"From Billy Budd's own grave," said another, answering his question.

"Grave robber!" they all began shouting.

One man began pummeling Bertie, when a tall man stepped into their midst and knocked the other man backwards. Soon a second man took his part, raising his quarterstaff to strike the tall man, but he parried the blow expertly, and quickly brought him to the ground with a leg trip without having to strike a blow.

"You have no cause to interfere here, Robin," said the man from on his back.

"No cause?" said Robin Hood, scornfully. "A pack of men bullying one small boy. I'll make that boy my cause any day."

"But you don't know the facts, Robin," said a large man, stepping from the crowd. "We caught him with Billy Budd's belt. He's a grave robber."

"That's sure! To be certain!" came the cries from the rest of the crowd.

"Now, boy," said Robin severely to Bertie, "where did you get this belt?"

"From Captain Hook," replied Bertie, thoroughly frightened. Bongo nodded his head, while still suspended by his tail, that Bertie was telling the truth.

"Who is Captain Hook? One of the Shire-reeve's men?" demanded Robin.

"No," said Bertie honestly, "a pirate in Never-neverland."

Scornful laughter arose from the men.

"You'll not lie to me, boy," said Robin angrily.

"I'm not lying! I haven't robbed anyone's grave!"

"We'll go and see then!" shouted one man in the crowd.

"Yes! Go and see!" shouted the rest of them. Someone pulled Bertie to his feet, and he was half-dragged after the mob of men.

"Wait!" cried one of the men.

They turned and paused.

"This may be a trap," he said. "The boy may have been sent to draw us out. After all, Billy Budd is buried on the Earl's jousting field."

While the mob hesitated, Robin spoke up. "Give the boy to my custody, and tonight, on my oath, I shall go with you, Hugh, to the graves of Billy Budd."

"Give me your hand on it then, Robin, and I shall meet you on the Edwinstowe Road at sunset. Until then."

Quickly, before a word of dissent could escape the crowd, Robin took Bertie's arm as Bongo was tossed to him, and the three stole silently away.

When they had gone a little ways, Robin doubled back upon their path to determine if they were being followed. Seeing no one, he said, "You can relax now, boy. What's your name?"

"Bertie," he said. "And this is Bongo."

"What is he?" asked Robin, eyeing him suspiciously.

"A monkey," replied Bertie. "They come from another country a long ways away. That's why no one knew what he was, but he's not a devil, although he does get me in trouble at times."

"Does he eat deer?" asked Robin.

"No," laughed Bertie.

"Then I don't think he'll do much harm in the forest," said Robin.

"Are you Robin Hood?" asked Bertie, who could not help betraying his awe.

"So I am called now," replied Robin, "but once was Robin of the Wood, because my father was the King's Forester, but he was murdered by the Shire-reeve."

"And are you the best shot in all England with a bow and arrow?" inquired Bertie.

"I don't know of anyone better," said Robin thoughtfully, "but that doesn't mean that I am the best."

"Well, I was telling Marian that you were," Bertie told Robin.

Robin's interest in Bertie perked up. "How do you know Lady Marian?" he asked.

"That is where I stay," responded Bertie.

"Then you must be a Norman relative of hers from the continent. We Saxons will have to show you more respect. I'll take you to her tree before more trouble comes your way," Robin said.

They walked along in silence for a time, Bongo riding on Bertie's back, when all of a sudden Bertie blurted out, "Robin, before you take me home, would you teach me how to shoot a bow and arrow?"

Robin stopped, somewhat flattered.

"It can't be learned in a day, y'know. It takes time and practice. And besides, you have to have the right bow."

"You could make one for me, Robin," pleaded Bertie.

Bongo tapped Bertie on the shoulder and pointed at himself, while nodding his head vigorously up and down.

"My monkey wants a bow too," added Bertie.

"Can he shoot?" asked Robin.

"Well, he can throw coconuts, so I guess he can shoot a bow and arrow," Bertie said.

"Very well," said Robin, "come along to the banks of the Idle, and there we shall find two young willow reeds for you to make you worthy of our Shirewood band."

On the way they came upon a clearing at the far end of which, one hundred fifty yards away, stood a buck with a tall stand of antlers. In one swift motion Robin unshouldered his bow and drew an arrow from his quiver. Notching the arrow in his bowstring he pulled it back all the way and released it, seemingly not pausing to aim. Flying with barely any arc, the arrow dipped into the buck's chest, and he fell to the ground, mortally stricken.

"Wow!" said Bertie.

"I know a poor potter and his wife who have not had meat in months, who will appreciate that buck," stated Robin. "Their tree is not far from here."

So overjoyed was the potter at the prospect of feeding his family, the heads of his many children peering at them from among the tree's roots, that he insisted upon giving Robin two of the finest products of his craft, drinking mugs which he filled with fermented honey and water. While Robin and Bertie drank the mead, his children played tag in the roots and lower limbs of their tree house, screaming with delight when Bongo would drop out of the tree onto their heads.

When they took their leave and Bertie stood up to go, his head buzzed as if a thousand bees had made a comb in his head. Robin chuckled, and hoisted him up on his back, whereupon Bongo sprang down from a branch on the tree onto Bertie's back. The potter's eight children stood watching and waving at Bongo till they were out of sight.

"It was your Christmas and birthdays all rolled into one," said the smiling potter to his children, and none of them as long as they lived ever forgot the long-tailed creature that played with them in the woods that day.

Even with the added weight of Bertie and Bongo upon his back, Robin's huge strides soon put them along a slowly meandering river.

"You see the river is in no hurry to get anywhere, so they call it Idle," said Robin, reflectively. "I'm like that too, nowhere to go—live day-to-day."

"But you could be a leader of men," protested Bertie.

"Lead them where?" said Robin, puzzled.

"Well, to a better life," Bertie persisted.

"Ah, no, not for me," said Robin.

Suddenly Robin stopped and let Bertie dismount from his back. Descending part way down the river bank, he cut a shaft of willow to make Bertie's bow; then he turned and looked over his shoulder at Bongo, who was watching him intently. Robin acknowledged Bongo's gaze, nodded, and cut another smaller willow wand. Returning to where Bertie and Bongo stood, he sat cross-legged on the bank and began fashioning their bows. At each end, he cut a notch, then from his pockets took long rolls of dried animal gut. Deftly he cut and tied the gut to each end of the bow, and handed the finished products to Bertie and Bongo.

"The little fellow's bow is about the size of my first one," said Robin.

Bongo stood gazing at his bow in rapt wonderment. Watching him, Bertie could not help smiling.

"Now," said Robin, "we must see how true they are."

Plucking an arrow from his quiver, he shot an arrow into a tree ten paces away. Then he took another arrow, fitted it into the bow, and showed Bertie how to hold and draw the bow. Bertie fired and missed the tree. Bongo held out his hand for an arrow, which was longer than Bongo's bow, and took it from Robin, who said,

"Aim at my arrow."

Bongo fired and Robin's arrow quartered as Bongo's arrow shattered the shaft. Bongo stood there holding his bow and looking as if he had done a terrible thing. Robin looked from Bongo to his shattered arrow, and then back again, as Bongo's face began to redden.

"I'm afraid he shows-off too much at times," said Bertie.

Robin soon discovered that Bertie was closing his eyes just before releasing his arrows, and that was why he was always missing. Then Robin told him the secret of his marksmanship.

"Before you release the arrow," instructed Robin, "*see* it flying to the target. Then you will have better luck."

Bertie tried it and hit the mark. "It works!" he yelled.

Then it was time for Robin to take them back to Marian's tree. As they walked along, Bertie said, "Robin, you promised the man to go with him tonight to the graves of Billy Budd. Why did you say *graves*?"

"Because he was quartered by the Norman knights," replied Robin, "and lies buried in four different graves. If you didn't know that I'm sure you didn't rob his grave."

"I didn't," replied Bertie, "but do you know which grave his belt is buried in?"

"I do," answered Robin.

"But if my belt is the same as *your* Billy Budd, but it came from another Billy Budd in another time, what does it mean?" asked Bertie.

"Ask Marian," said Robin. "She knows many things that seem to make no sense."

Having walked most of the morning, Bertie was ready to rest when Robin said, "Ahead is a place where the Shire-reeve's men sometimes try to waylay forest folk. If I climb to the top of this tree, I shall be able to see whether or not the way is clear."

At Robin's insistence, Bertie and Bongo climbed into the lower branches of the tree to be out of sight, while Robin scanned the forest ahead. Bongo having scurried up in no time at all, Robin said, "Your little friend could give you a lesson in climbing as well as archery."

Bongo looked smug.

When Robin had descended, he reported no danger ahead, and they set out walking once more.

"I often wonder," said Robin, "about where lies the end of the world. And if a tree were tall enough, could one see that far."

Bertie laughed. "Of course not, because the earth is round. No matter how far up you were, you could only see half of the earth."

"What's that you say?" said Robin, looking alarmed. "The earth is *round?*"

"Yes, it is," replied Bertie.

Robin thought a while and said, "But it is flat."

"It only seems to be flat," countered Bertie, "because we can only see a little ways.

Look, I can prove it to you," he said, making his hands into two fists. "Here is the earth, and here is the sun. Earth spins on its axis, and when directly facing the sun, it is warmest, then it gets colder in autumn, and coldest when tilted away from the sun. Don't you see, the seasons are different because of the difference in the earth's relation to the sun."

"No," said Robin, "that's just nature. It's always cold in Januarius."

Just a few hours ago, Robin had been a hero in Bertie's eyes, but now he seemed too stupid for words. As far as Bongo was concerned, he didn't care whether or not Robin knew that the earth tilted on its axis. So he rode on Robin's shoulder the rest of the way home to Marian's, with Bertie lagging behind.

Just before they arrived, Robin said to Bertie, "Do you have the writing, then?"

"Of course," said Bertie. He was about to add, "Do you think I'm as stupid as you are?" but held his tongue.

"I have a rhyme in mind," Robin said, shyly. "If you could set it down for me on paper, I would be obliged."

"Then what do I do with it?" inquired Bertie.

"Give it to the Lady Marian," replied Robin.

"What?!" exclaimed Bertie. "Why don't you just say it to her, if it's for her."

"Oh, I could nay do that," said Robin. "I'd sooner fight all the Shire-reeve's men barehanded."

"All right," said Bertie, peevishly. "If you made a bow and arrow for me, I guess I can write down your silly rhyme."

Out of his pocket he took a scrap of paper and a stub of pencil and waited for Robin, who recited as follows:

"Bitter the arrow of you
That strikes the unarmored heart.
Bitter me, who has in my heart you.
The arrow kills with its heartwound,
Love's arrow does not, though I wish
It did with all my dying heart."

When he had finished writing, Bertie stuffed the paper in his pocket. To him it was as if Robin had betrayed the admiration that Bertie felt for him by revealing his affection for Marian. For his part, Robin was embarrassed by the revelation of his rhyme, and so neither one said a word again until they had reached Marian's tree.

She was sitting outside with her good side, as Bertie came to call it, facing them. The sunlight made her hair fairer, and made her skin seem a pale gold. Bertie stole a glance at Robin, for he was afraid that in that light she would seem even more enchanting to him. Certainly now he looked like a lovesick calf. Bertie was disgusted with him.

Hearing their footsteps, Marian turned to look upon them, and the sudden contrast of her dark side was shocking to see. For a moment Bertie hesitated, and then he rushed forward to her.

"Oh, Marian, some men chased us and Robin saved us, and made us bows, and he shot a deer and gave it to some poor folk, and we've had such adventures!"

Now, when he was with Marian, Robin was his hero again, and he showed Marian his bow, and Bongo extended his little bow for her to see.

"Robin," said Marian in curt greeting.

Robin doffed his Lincoln green cap and bowed low to her. "Lady Marian," he said.

"I would invite you to stay for tea, but I have only been waiting for these two to return. We must go to Nottingham Castle at once."

"But the King is there now," said Robin.

"I know," replied Marian. "Queen Eleanor has invited me to dine with her, and I cannot leave these urchins alone."

"They could stay with me," responded Robin.

"No, they had best be with me."

"But, how are you to go? You cannot be there before nightfall, and the Shire-reeve's men may take you."

"Not while I have this," said Marian, holding up the badge of safe-conduct sent by Eleanor of Aquitane.

"I myself am meeting Hugh of Hawton to go to Nottingham tonight," said Robin.

"What business do you have there?" inquired Marian, eyeing him suspiciously.

"The matter of the boy's belt," answered Robin. "You have seen it, I am sure. When the men near Rimely saw him, they took him for a grave robber. They let him go on my word that I would go with Hugh and see if the belt of Billy Budd still lies with its owner."

"You will find it there," said Marian.

"Then how comes the boy to have it? He wants to know himself, but I could not answer," said Robin.

"The belt moves on to other times," Marian replied. "The boy claimed it in another time, and now has come to our time."

"But what does it mean?" said Bertie.

"It means," said Marian, "that the wearer of the belt is a force for Light, who has earned the right to wear it. In our time Billy Budd was our only savior, but he met a bitter fate, as is the wont of saviors in the world. What happened to him in your time?"

"He was hanged," Bertie replied.

"A sudden end," said Marian, wryly. "But you can now be the force for good that Billy Budd was for us. You have a purpose, do you not?"

"Yes," said Bertie, mildly.

"Then you and I together shall bend the King's ear to our purpose."

Marian had consented to their being accompanied by Robin and Hugh of Hawton.

Several times during the journey, Bertie started to remove Robin's rhyme from his pocket to read it to Marian, but each time Robin stopped him. Finally in exasperation, Robin took Bertie aside and told him, "Read it to her when I am gone!"

Bertie nodded, and the rest of the journey was uneventful, except that when Hugh of Hawton learned that Marian would be in the King's presence, he recited a litany of grievances of the forest folk against the Shire-reeve, which he hoped—with all due respect—Marian would pass on to the King.

When they were within sight of the castle, Robin and Hugh took leave of them to go to the eastern end of the jousting field containing one of the four graves of Billy Budd, wherein lay the fabulous belt of the serpent swallowing his tail.

"Is this the King's castle?" asked Bertie.

"No," replied Marian, "but it was built by his grandfather, William the Conqueror. Now King Henry only stays here when he comes to administer justice in these parts."

"Is he a just king, Marian?" Bertie inquired.

"Very just. But before the night is out I shall give him an earful about the Shire-reeve of Nottingham."

"What will he do about him?"

"That remains to be seen, but the King is a just man, and I think he will be displeased surely."

Just then two riders approached from the castle, challenging them, but when Marian showed Queen Eleanor's safe-conduct badge, one of the riders dismounted, and they were given his mount to ride in triumph to the castle, where the gates were thrown open for them.

"Have you ever dined with a king?" Marian asked Bertie, as they were ushered towards the banquet hall.

"No," he said, "I've never even seen a king before—except in the movies."

"Give him and Queen Eleanor the same respect you would your own parents, and you will not go far wrong. Keep your eyes upon your plate and do not stare at them. And if the king and queen argue, take no notice."

"Do they argue, Marian?"

"Sometimes, for he cannot forgive that she was once the wife of his mortal enemy the King of France."

"Then how did she come to be the King's wife?" asked Bertie in amazement.

"Oh, too many questions!" exclaimed Marian impatiently.

They were now outside the banquet hall, the doors of which were thrown open at their approach. Since none of the King's servants knew them, they could not be announced; hence their entrance was accompanied by great confusion, particularly because of Bongo's presence.

At the head of the banquet hall sat King Henry II, a man of 30. On his left was Queen Eleanor, a striking woman eleven years his senior. On the King's right sat Thomas a Becket, Archbishop of Canterbury, the highest ranking prelate in all of

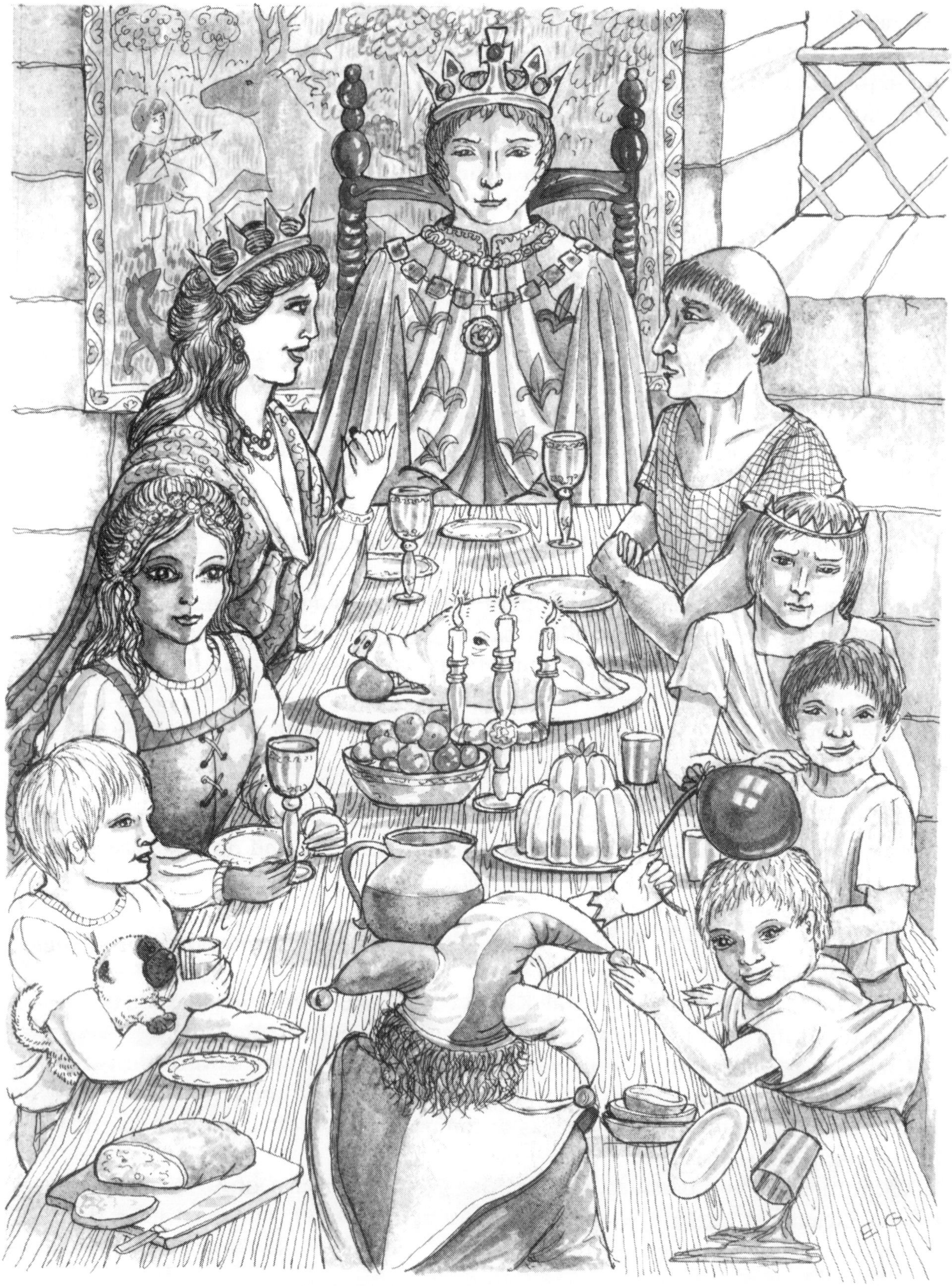

England. Despite his lofty status, he wore a monk's common habit and a hair shirt under his garb. Also present were the King's sons, Prince Henry, 8, Richard, 6, and Geoffrey, 5.

Maid Marian bowed low before approaching the table, and whispered to Bertie to do the same. At the same moment, the King's Fool left the foot of the table and bowed low to Bertie as Bertie bowed to the King. Then clasping Bertie's hands in his own, the Fool began to dance him about the banquet room, much to the amusement of all save Bertie.

Henry: Ah, Maid Marian, you have lost your maidenhood and brought forth a child.

Marian: No, my liege, but someone to pique your infinite curiosity about the world.

Henry: Indeed. And how can one so young be of interest? Is he well-travelled?

Marian: Well-travelled, indeed, your majesty, for he comes to us from the 20th century.

Henry: And how comes he from the 20th century?

Fool: On his ass like Christ.

Becket: (Humorously) By the grace of God, I should think.

Henry: Maid Marian, will you take a place next to my Lady, for until you arrived she has been not a queen but a lady-in-waiting.

And my young guest, would you take a place below an archbishop? I would not, and that is why we have been quarreling.

Becket: If it comes to quarrels, Henry, then I choose young Robin Hood for my champion.

Bertie looked perplexedly across the table to Marian.

Marian: (to Bertie) A quarrel is another name for an arrow.

Bertie nodded that he understood.

Henry: Does our young traveller have a name?

Bertie: Bertram, your majesty. But you may call me Bertie. And my friend is Bongo.

This remark set the King and everyone to laughing.

Henry: And this is my Queen, Eleanor; my sons Henry, Richard, and Geoffrey; and my Archbishop, Thomas a Becket, in his sackcloth and ashes phase. When I last saw but a few months ago, he was in ermine robes and laid the most sumptuous table in all of Christendom.

Becket: That was before you appointed me Archbishop, your majesty.

Henry: And should the Archbishop be less splendid than before?

Eleanor: Yes, Thomas, you must tell us why this change came over you?

Becket: There is no great mystery, your majesty, simply a realization that poverty is closer to the way of our Lord.

Eleanor: Is it true that you nightly wash the feet of thirteen beggars?

Becket: Yes, I find it helps keep my lofty position in perspective.

Henry: Perhaps tonight, Thomas, you would consent instead to give my Fool one good wash from head to toe, for his black humor stinks surely.

Fool: Scrub a fool and find a king under the skin.

Henry: Had you said the obverse of that proposition, I would have you stocked, Fool.

Fool: Better fools in stock than out of stock.

Henry: Unless their wit be in short supply.

Eleanor: Your pet, Bongo, what kind of animal is he?

Bertie: A monkey, your majesty.

Eleanor: So like a human, except of course for the tail. Have my boys seen one of these before?

Geoffrey, Richard, Henry: (together) No, mother.

Henry: I have seen them in Spain. They come from the land south of the Mediterranean.

Eleanor: (to Bertie) Is he useful to you?

Bertie: O, very. He saved my life in our last adventure.

Eleanor: You see, Henry, you would be better protected by a troop of monkeys than your knights.

Henry: But can they ride horseback?

Becket: If you say truly that you come from the future, give us some morsel of knowledge of which we are ignorant in this century.

Bertie: (thinking) All right, the main misconception is that the sun moves around the earth.

Henry: Does it not?

Bertie: No!

Becket: But all eyes say nay, and see its daily journey through the sky.

Bertie: So it *appears*. But if you were standing on a rotating globe—as you are—then the sun would appear to rise and set. But it is the annual orbit of the earth around the sun which makes the seasons.

Henry: How is it then that we feel not the earth's movement?

Eleanor: You are of a scientific mind, then, even so young?

Bertie: I want to be a physicist when I grow up.

Eleanor: What, pray tell, is that?

Bertie: One who studies the physical laws governing the universe and formulates theories about the origins of the universe.

Becket: And are there none in your time who study the governing spiritual laws, God's law?

Bertie: Yes, there are still churches.

Becket: Then you were not educated in true doctrine, for when you say the earth moves around the sun, you blaspheme and speak a heresy contrary to Holy Scripture.

Bertie: That is what they told Galileo in 1615 when he tried to prove to them that they were wrong.

Henry: Ah, you have knowledge of the seventeenth century as well.

Bertie: Yes.

Henry: What say you to the archbishop's charge of heresy?

Bertie: I *know* that what I say is true.

Becket: And if you are told that what you say is contrary to Holy Scripture, and that you must not advocate or believe it, even if you think it to be true, what say you then?

Bertie: I will say then what Galileo said to the Pope and the court of Inquisition which tried him. "You cannot hold my mind in check, for it flies to truth as a dove to heaven."

Eleanor: Who was this Galileo?

Bertie: An Italian, and one of the first great scientists of the world. He was not speculating when he said the earth moved around the sun. He had built a telescope which proved it.

Henry: And what is a telescope?

Bertie: A series of glass lenses and mirrors for clearly seeing distant objects. When Galileo asked the Pope's representatives to look through it, they said it was a trick of the Devil.

Becket: And most certainly it was, for whatever is contrary to Holy Scripture and would have us believe otherwise is manifestly the Devil's work.

Bertie: No, that is why science is more important than religion in my century, because people will not accept dogma over truth.

Becket: Your majesty, let me take the boy in hand and see that he is educated by one of my monks, purged of his delusions before his immortal soul is lost.

Marian: The boy stays with me. You have no claim to him.

Becket: What say you, your majesty?

Henry: I find his mind works much like mine and would be loath to curb it, even for the sake of his soul.

Bertie: My soul is my own.

Silence now prevailed at the banquet table. Becket felt rebuffed, and King Henry did not want to aggravate him by speaking further on the subject. On their part, both Eleanor and Marian feared that Becket would pursue the matter. To change the subject, Eleanor spoke up.

"With your future knowledge, do you know the succession of English kings?" she inquired of Bertie.

"Ah, yes," said Henry, "a very wise question with which to test him. Who came before me on the throne, and who before him?"

"Stephen, and Henry the First," replied Bertie brightly.

"Brava!" exclaimed Eleanor, applauding. "How very bright you are. I only wish my sons would show as much inclination towards learning, and less towards fighting."

"They are young," responded the King, "and if still uncultured pearls, they are yet within the half-shell of your influence."

"My liege," said Eleanor smiling, and bowing her head slightly.

"Stephen is best forgotten," said King Henry, "an unpleasant interlude between the good works of my father and my own good intents.

"I do not know what history will say of me," continued the King to Bertie, "but I shall share with you my plan for England—law, order, justice, for all men. And so that there will not be bad blood and rivalry between my sons for my kingdom, I intend to apportion it while I am still alive, so that there can be no question about my wishes and so that no son is disenfranchised.

"In seven years, therefore, when he is 15, I shall crown Henry so that he may rule jointly with me until my death. To Richard, I shall give the French province of the Aquitane, which came to me in marriage with Queen Eleanor. And to Geofrey I shall cede all of Brittany in France. Those are the true intents of my will. Now tell me who succeeds me as king, if not Henry?"

Bertie thought for a moment and then said, "Richard is next in 1189."

The King frowned and said, "You are only a commoner, boy, but you have dealt a death sentence to a king."

Bertie looked perplexed. Eleanor said to him, "You gave the King the year of his death—hence, a death sentence."

Bertie was about to make apologies when the King blurted out, somewhat angrily, "Who follows Richard then?!"

"John," said Bertie quietly.

"Who?!" roared the King.

"John," spoke up Bertie.

"Then he is not of my line," replied the King sadly.

"But he is," affirmed Bertie.

"I am now forty-one," Eleanor said to Bertie, "and my child-bearing days nearly at an end. It seems unlikely that I shall bear another child. Does this mean that Henry will wed again?"

"Why put credence in a child's prattle?" said Becket. "The future is known only by God, and to encourage the boy in this is to encourage more blasphemy!"

"Am I to be king after you, father?" asked Richard.

"The Archbishop has put an end to our inquiry," said Eleanor coldly.

"Henry, a small point perhaps, but it requires clarification," said Becket to the King.

"Yes, Thomas?"

"You said *you* would crown young Henry king. Surely you recognize that the powers of investiture lie with my office as Archbishop. You can designate young Henry to be king, but only I may crown him."

"That power was given away by Stephen," replied Henry. "My father, however, claimed it."

"And Archbishop Anselm opposed him," countered Becket. "You have in mind to once again limit the Church's authority. I know this to be true because my bishops tell me that a council is to be held next year at Clarendon. Why have you not raised these matters with me?"

"Because," replied the King, "you have become increasingly reluctant to curb Roman power in this land."

"And what 'curbs' do you now intend?" questioned Becket.

"Think of them instead as reins upon a wild charger," said Henry, "that with the bit in his teeth, and goaded on by the Pope, tramples all who would seek to restrain him."

Becket smiled slightly and said, "Your analogy runs unbridled away from my question."

"How acute of you, Thomas," responded King Henry. "It is a matter which I have raised before and it still catches in my craw: abbots and clerics above the law of the land and beholden only to Rome."

"And cannot the Pope judge them as severely as a king?"

"Evidently not," the King answered, "for they go unpunished for crimes which would send a villein to death."

"What do you hope to accomplish at Clarendon?" questioned Becket.

"I shall present a list of the ancient customs of the realm, which I should like the clergy to observe. Further, I shall reassert the royal right to elect bishops. And with regard to clerics accused of crimes, I shall require that if they are found guilty when tried by an ecclesiastical court, they shall be turned over to secular justices for punishment. In other words, they will no longer be above the law of the land."

"I would refuse to sign such a document, or to affix the seal of my office to it," said Becket.

"Then I would remove you from your office and make another archbishop."

"You cannot," replied Becket. "I answer only to Rome."

"That is exactly the pith of our disagreement," roared the King, slamming his fist down on the table. "You are like a worm boring insidiously into the apple of our England, and rotten priests are at the core of the matter. Confound it, Thomas, I have made my name as a fair man, whether dealing with baron or earl, freeman or villein. I cannot in good conscience turn my back upon crimes committed by the clergy. No one can be above the law in my England."

"Your law or God's law," replied Becket. "The bishops will move over the chessboard of Clarendon as my hand directs them, and my hand is moved by the Pope."

"An ill-advised metaphor," said the King, "for in the game of chess there is no pope, and ultimate power resides with the king. I am already assured of the support of enough bishops to pass articles at Clarendon subjecting clerics to punishment in secular courts."

"Nevertheless," replied Becket, "your stratagem will ultimately fail. I know the Pope's mind on this. He will excommunicate the bishops, and yourself very probably, and withhold mass and all holy rites in England. The fear of hell in your subjects will overwhelm their allegiance to you."

"Cosmic blackmail!" exclaimed the King.

"Call it what you will; you will end by eating humble pie, and I would relish seeing that no more than you would eating it."

"In the Aquitaine," said Queen Eleanor, "bickering at table is regarded as the worst of bad manners."

Angered at her intrusion into his attempt to settle matters with Becket, Henry replied angrily, "You left France and the Aquitaine when you left King Louis' table and his bed!"

"You proceed from the coarse to the coarser," replied Eleanor, coldly.

"Pray then," said the Fool, "let us have the next course."

Laughing heartily, Henry shouted, "And so we shall! Lemon trifle!"

A servant standing over his shoulder clapped his hands, and a small army of servants entered the hall, clearing dishes, whisking away crumbs, mopping up spills. They were accompanied by tumblers and jugglers, juggling balls and knives, and one of exceptional skill who tossed fiery brands into the air.

A wooden ring illuminated by candles was set before the King, wherein was placed the lemon trifle, a pastry dessert.

"Do you still have lemon trifle in your day?" Eleanor asked Bertie, whose eyes and the eyes on the young princes had grown very large.

"What is it?" asked Bertie.

"A pastry topped by egg-whites sweetened with lemons from Spain. The pastry is layered with honey," said Eleanor.

"I don't think so," responded Bertie. "But if I ever get back I'll ask my mother to make it."

"And you can tell her," said Marian, "that the recipe was given to you by the Queen of England."

Everyone laughed, and the King spoke to the servant at his shoulder. "More wine for my guests, and more water for the Archbishop to whet his appetite for argument."

"And more claret," called the Fool, "to muddle the clarity of Clarendon."

"An ill-advised pun, Fool," said the King. "Tell me, young man," he said to Bertie, "do they still pun in the England of your time?"

"Oh, yes, your majesty, outrageously."

Everyone laughed.

"And it was a pun," continued Bertie, "that got me into trouble in my very first adventure."

"How so?" inquired the King.

"I called the Toad, Pigeon Toad," Bertie replied.

"By toad you mean some low fellow?" inquired Eleanor.

"No," said Bertie, "an actual talking toad who stole and was put in jail."

"No doubt one of Lady Marian's familiars," said the King, laughing.

"If you would be familiar with Lady Marian, you'd best not be at moonlight," said the Fool.

"Henry," said Eleanor, "your insinuations about witchcraft to Lady Marian are in even poorer taste than your remarks to the Archbishop."

"I have already forgiven him, your majesty," said Becket. "We have long been friends as well."

"You served me well as Chancellor," responded the King. Standing, the King cut the pastry into equal portions, and it was served to all those at the table. Looking at Bertie, he said, "When one is king it is as if every day were your birthday."

They ate, and the table was cleared once more, and wine poured. Musicians entered playing harps and dulcimers.

"Richard, why don't you take Bertie out to see your ponies," said Eleanor.

"Come on, Bertie," said Richard, leaping up excitedly. For a time there was a cold silence at the table, then young Henry asked, "Are you going to give Richard my kingdom, father?"

"By heaven no!" roared the King. "Thomas is right. This young soothsayer sows the seeds of dissent."

The Fool jumped up and did a quick little dance. "King Lear divided his kingdom into three parts, just like Gaul," he sang in sing-song, "and he found the castles' doors barred wherever he went."

"And," roared the King, "if I divide my Fool into parts, then my similarity to Lear shall end there."

"Calm yourself, Henry," said Becket, placing a restraining hand upon his arm, which had gone to the hilt of his sword. "The Devil's ways create a tempest in a teapot."

"He is not of the Devil's issue," said Marian angrily, defending Bertie. "He had seen my golden thimble somewhere in the future and knew the inscription within it while it still rested on my hand."

"But your name is Marian, is it not?" questioned Becket. "And was your name not inscribed within? He need learn only your name to know the inscription."

Marian made no reply, and while King Henry and Becket conversed, Eleanor spoke softly to Marian.

Presently Bertie ran in bleeding, pursued by Richard, whom Eleanor grabbed as Bertie sheltered behind Marian. Richard was carrying a wooden sword, with which he had struck Bertie in the head. Marian blotted his blood with her handkerchief.

"There's a lion heart!" cried the King to his son.

"He doesn't fight fair!" exclaimed Bertie.

"There now, it's nothing," said Marian, soothing Bertie.

"He'll fight against you with the French king!" Bertie cried out defiantly to the King.

"We had best separate these boys," said Eleanor diplomatically. "You'll stay the night," she said to Marian, "and then in the morning you'll be escorted by horseback

back to the Shirewood."

"We wouldn't want Robin Hood to waylay you," said the King.

"Robin Hood is our protector," replied Marian, indignantly.

"What?! That outlaw?!" exclaimed Becket.

"If it were not for the band in Shirewood," said Marian, "poor folk such as myself would all be in our graves now."

"And how comes it that the Earl of Nottingham protects you not?" inquired the King.

"When our lord is in his castle and not upon the continent, he knows naught of the plight of the people. His Shire-reeve does what he will with us, and we are in hiding in Shirewood."

"Know you of this, Thomas?" King Henry asked Becket.

"Only that several of my clerics have been relieved of their purses while on their way to Lincoln cathedral," replied Becket.

"The Church cannot quarrel with his example," said the King, smiling.

"How say you?" asked Becket.

"He takes the fat purse of the cleric and doles it out to the poor. Is that not Christ-like, Thomas? 'Twas only recently that your own display of wealth was second only to my own."

Becket said nothing.

"Tell me, Maid Marian," inquired the King, "why my folk must shelter in the wood when I have brought them the protection of English Law?"

"Laws are upheld by men or broken by men," replied Marian. "The Earl's *cnihts* have behaved towards us like common criminals, your majesty."

"I see that you use the Saxon word for knight, and not the French *chevalier*. Why, pray tell?"

"Chivalry is implied in the French meaning, your majesty, and the conduct of your knights is chivalrous, and so I use the French in referring to them, but the knights of the Earl of Nottingham behave no better than servants, and so I use the Saxon *cniht*, meaning servant or retainer."

"This is the first imputation against the Earl's knights to reach my ear, but like the bee homing to the hive, you may be sure that it makes honey there. Tell me of the conduct of these knights?"

Marian looked uneasily at Eleanor, who touched her hand reassuringly.

"Henry," said Becket, "the hour is late for one who is called to prayers at five before the dawning, so I beg to take my leave. The misconduct of the Earl's knights is not an ecclesiastical concern, fortunately for me, or you would have yet another cause for your Council at Clarendon."

"Goodnight, Thomas, and may you find rest tonight despite your hair shirt."

When Becket had departed, the King turned to Marian and asked her to continue.

"Pursuing your metaphor, your majesty, the only beeline to your ear heretofore has been through the shire court, presided over by the Earl of Nottingham and the Shire-reeve. The people of Shirewood are mostly freemen disenfranchised by the fiefs of our land given to the Earl's knights. Robin Hood is really Robin o' the Wood, so-called because he is the son of the King's Forester for Nottingham. But his father was killed by the Shire-reeve, and he has been in hiding ever since then. If he has taken to robbing to survive, it is no one's fault but the Shire-reeve's.

"My own story Queen Eleanor knows well enough. I am half-Saxon, half-Norman, half-fair, half-dark, as my complexion bears witness. I am, therefore, half-slave, half-conqueror, but my heart flies to the oppressed, and so I have chosen to join with the Shirewood's people. Rebellion is not a happy state, and our quarrel is not with our king, but with his minions.

"When our last pleadings fell on deaf ears at the shire court, the Shire-reeve whispered to the Earl, and subsequently we were offered full pardon if we could find a champion who would represent our cause in mortal combat—a joust by the French rules and standards."

"A joust!" exclaimed the King. "That cannot take the place of my English law! Not on English soil! That is a French affair. What right have they to resurrect it here?!"

"None, except that *they* are making the laws in your absence," replied Marian.

"They!" roared the King. "Who are *they*?!"

"The Earl of Nottingham, and the Shire-reeve," answered Marian. "But the knights are a law unto themselves. They take what they want from freemen and villeins alike—a farmer's pig, or his daughter's maidenhead. And those who resist are killed."

"Anarchy afoot in Nottingham!" cried King Henry. "This serpent shall be squelched!"

"Do save your ire for the Earl, Henry, and let Marian go on," said Eleanor.

The King sat back in his chair and remained silent, only the high color in his face betraying his mood.

"And so you sought a champion," said the King, at last calm again.

"Was it Robin Hood?" Richard blurted out.

"Richard, don't interrupt your father," admonished Eleanor.

"No, Robin was merely a boy. We found no one to champion us, for the Earl and the Shire-reeve had declared the Black Knight would take their part, and the Black Knight had killed many men in jousting."

The Fool cut a caper and said, "Just as a Fool kills many men in jesting by the sword of his wit."

"Sit down, Fool, thy wit is a tired thing tonight," said the King. "I have heard of the Black Knight on the continent," said Henry, "but were he to kill here on English soil he would answer to English law."

"Then your majesty will not be pleased by the rest of my history," replied Marian.

"Do go on," said King Henry.

"There was a poor villein living in Edwinstowe," said Marian, "who one day heard of our plight and came forth to champion us. Almost certainly he would be giving

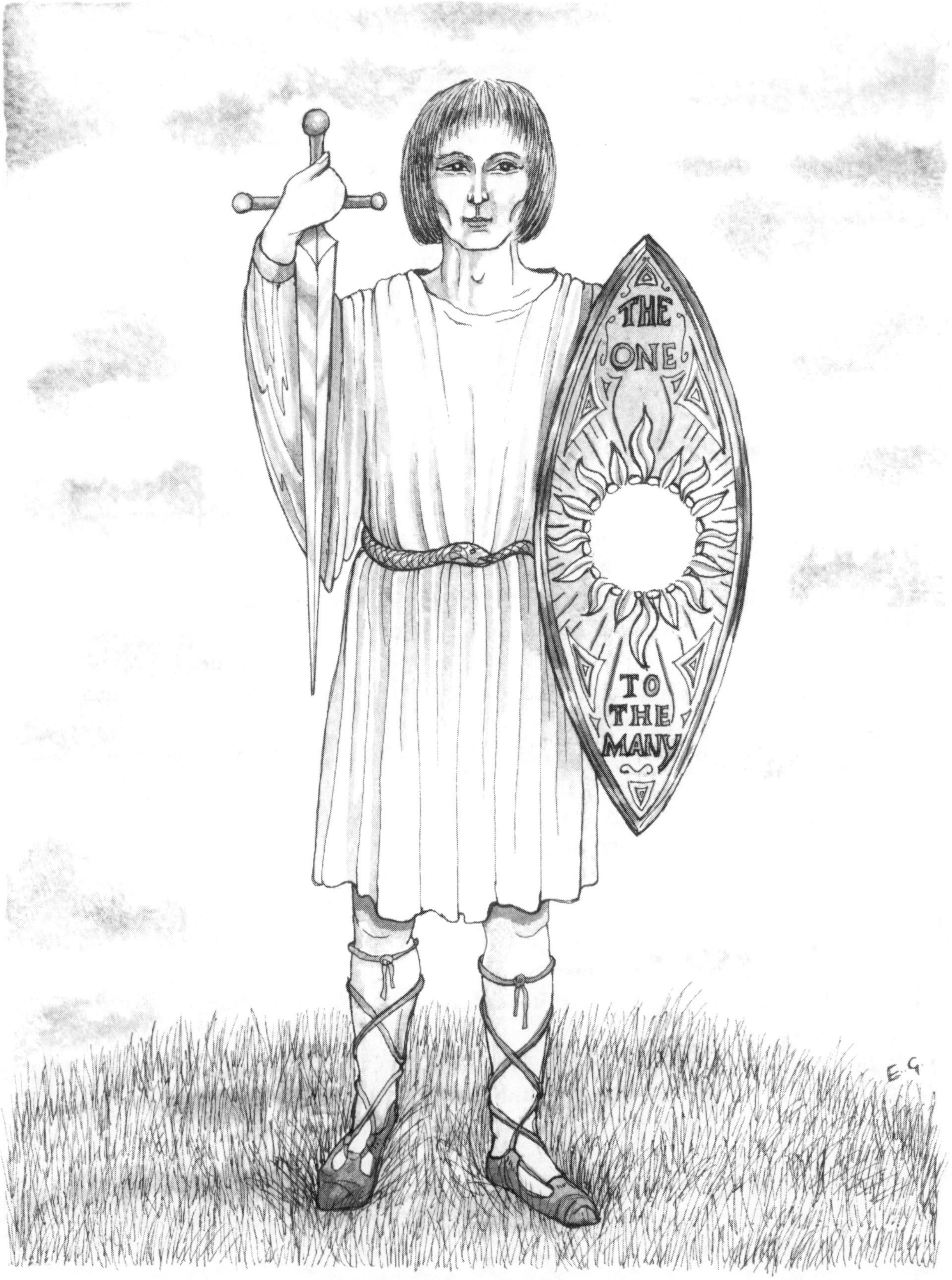
THE
ONE
TO
THE
MANY
E.G

his life, because no one could match the skills of the Black Knight. Yet he was cheerful and uncomplaining, almost simple-minded in regard to the dangers facing him. In his village he was known as one who shared his life with others, whether it was his own meager bread or the labor of his strong body. The people of the shire gravitated to him now, as to a shining star, and he in turn radiated light as if infused with the spirit of his sacrifice.

"For him the Smith forged a shield upon which was painted a many-rayed sun of golden light, and beneath it a motto: 'The One to the Many,' which, as I took it, meant his life given to the people.

"But the men of the wood trained him well, using trunks of trees for lances, each day progressively heavier, until he was able to wield the heaviest with the greatest dexterity.

"Then the day came for his encounter with the Black Knight. He showed no signs of fear, no regret for his choice to champion our cause. The joust was held in Nottingham, and Norman knights and lords from many shires afar, journeyed here for what they thought would be simple slaughter for the Black Knight. But his lance shattered on our champion's sun shield, and he was unseated. Our champion could have run him through with his lance, but scorned the unequal combat and himself dismounted to do battle by broadsword.

"For many hours they fought, and the wounds of both were terrible indeed, but just before sunset our champion struck a fatal blow and the Black Knight sank to the earth. The Norman earls and their ladies fell silent, and a pall passed over the gathered throng of knights like the shadow of the Angel of Death, while a great cry went up to the sky from the Shire's people for our deliverance.

"But it was not to be. The promise of amnesty and an end to our persecution with the prevalence of our champion was a cruel lie; for four knights disengaged themselves from their ranks and set upon our champion from all sides. Already exhausted, he was no match for them and soon fell.

"But then the real horror began. His body was quartered by the knights and dragged to the four ends of the field of combat, and his entrails scattered to the winds. That night by a bilious half moon, we returned to the earth what remained of our champion, Billy Budd."

The King was silent for a long time, and then spoke. "The atrocity you have just recounted will not go unpunished. The Earl of Nottingham, the Shire-reeve, and the four knights will be brought to trial, if what you have told me is the truth."

"It is the truth, every word," replied Marian.

That night, as he slept in the chambers provided for him and Marian by the King, Bertie had a dream or vision, for he saw a tall, white-bearded man standing in the room, leaning on a staff, and was surprised to see his own still sleeping form on the bed as he stood before the man. He thought a question, "Who are you?" and the answer came into his head instantaneously: "I am Merlin, an aggregate of consciousness thousands of years old. Now I have incarnated as a fool, and wisdom wears the cloak of foolishness. I shall come to you again."

And with that he was gone. When he awakened in the morning, Bertie was not sure whether he had been awake or dreaming.

After Bertie and Marian had dined at breakfast with Queen Eleanor on pork sausage and loganberry pancakes smothered in butter and honey, and the Queen had departed, saying she would send her escort back to Shirewood with them, a knock came at the door. Thinking it to be the men of the escort, Bertie opened the door. There stood the King's Fool. Dumbfounded, Bertie said nothing as the Fool smiled at him.

"Who is it?" called Marian from within.

"Who, indeed," replied the Fool, leaning his head in the doorway and whispering to Bertie. "H-O-O spells Hoo. What makes one man good and another evil?"

"I don't know," said Bertie laughing.

"A letter," replied the Fool.

Bertie laughed again.

"H-O-O-D spells Hood. Robin Hood. Good or bad?"

"Good," replied Bertie.

"D-E-F!H-O-O-F spells Hoof. The Devil you say. Good or bad?"

"Bad," said Bertie, laughing again.

"G-H-I-K!H-O-O-K spells Hook. Good or bad?"

"Bad," said Bertie. "A pirate. But how did you know that?"

"All in the alphabet," replied the Fool. "Look at the stars. Think how many worlds there are."

"Millions," replied Bertie.

"How many alphabets?" asked the Fool

"Millions," Bertie again replied.

"Infinite. New stars always being born, new life, new alphabets," said the Fool, rhapsodically.

Marian came to the door and said, "Oh, it's you Nilrem."

"Why is he called Nilrem?" asked Bertie.

"The name means 'no remembrance.' He cannot remember anything from one day to the next. Some say he cannot remember from one moment to the next. Why are you here, Nilrem?"

"To riddle you," replied the Fool, "if you can remember it."

"We shall remember. Try us."

And the Fool recited the following riddle:

"To the supple woman
Marry the upright man.
Temper the fiery metal
With the balmy yew.
Resolve the pairs of opposites,

And so shall have you
Death in life, life in death,
Tempered to try their mettle."

He recited it once more, and then was gone. Marian then recited it with Bertie and when they were through, she asked, "Can you remember it?"

"Yes, replied Bertie, "but what does it mean?"

"That is for us to discover," replied Marian.

At that moment the men of the King's escort appeared at the door, and they were led down to the castle's courtyard where two horses were saddled for them, along with an escort of six knights. Bongo rode in the same saddle with Bertie, sitting in front of him. At first he seemed afraid of the horses, but later wanted to hold the reins and see the horses gallop.

As their horses walked into the first leafy shadows of the Shirewood, a bird call sounded in the distance, causing Marian to become at once alert. Overhead, on an oak bough, Robin and Hugh of Hawton had draped themselves as silently as snakes, awaiting a signal from Marian if their aid were required. But she brought out her handkerchief, pretending to wipe her face, and the column passed unmolested under the bough.

When they were once more safely within the roots of Marian's tree house, Bertie could not wait to question Marian.

"Why did you praise Robin to the King, when before you only said bad things about him?"

"That is the way of women," said Marian.

"But do you like him or don't you?" persisted Bertie, very perplexed.

"What does it matter?" asked Marian.

"Well, it matters because I think Robin likes you," said Bertie. "He gave me this to give to you before we left, but I didn't have time to show it to you."

Marian read Robin's poem, and smiled sweetly.

"You do like him!" exclaimed Bertie.

"I didn't know Robin could write," Marian said slyly, "but it's a pity he can't spell."

"Why do you say that?" Bertie asked, scrutinizing the poem.

"He asked you to write it down for him, didn't he?" queried Marian.

"Yes," confessed Bertie. "But how did you know that?"

"Why don't you read it to me," said Marian, "and I'll close my eyes and pretend Robin is here."

Bertie blushed at the thought of reading a mushy love poem. It was one thing to write it down as a favor to Robin, but quite another to have to read it. Nevertheless, when he saw that Marian had closed her eyes and was waiting, he read aloud as follows:

"Bitter the arrow of you
That strikes the unarmored heart.

Bitter me, who has in my heart you.
The arrow kills with its heartwound,
Love's arrow does not though I wish
It did with all my dying heart."

"Thank you, Bertie," said Marian, opening her eyes. "Robin has turned a rhyme in the tradition of courtly love, of which the French know more than the English, but perhaps Cupid's arrow has flown all the way from France."

Bertie blushed again and looked down at the ground. Marian came over to the table and pointed at the first line of the poem.

"Did you not know that there is a tree named 'yew,' and that arrows are made from its wood?" Marian inquired.

"I think so," replied Bertie.

"Well, Robin knows that," said Marian. "He makes all his bows and arrows of yew wood. Now look here," she said, again indicating the first line, "you should have written 'Bitter the arrow of yew.' Robin has made a rhyme of yew and you. He says he has me in his heart like an arrow of yew. '*Amor*' is the French word for 'love', so when he says 'the unarmored heart' it has the double meaning of a heart that has never loved before. It's quite clever really, even if it's his first. Did he say he had written any more?"

"You don't want him to write more of that mush?" Bertie stated disdainfully.

"I wouldn't discourage him," replied Marian.

Outside they heard approaching footsteps. Bongo had climbed onto the table to see Robin's handiwork, and when Marian warned that they should be silent and not move, he was caught with one leg in the air. Bertie began giggling because Bongo looked so ridiculous, and Marian was about to rebuke him when there came Robin's call.

"Hallo, Marian!"

Marian stepped outside to greet Robin and Hugh of Hawton. The way that she looked at him, Robin knew that she had read his rhyme, and he could not bear to look her in the eye. Mercifully she turned her gaze to Hugh of Hawton.

"We bring news of the dead, Lady," he said. "The serpent of Billy Budd still burrows in the bones of his bosom."

"Ah, I knew it would," replied Marian.

"Then there are two identical belts," stated Robin.

"There is only one belt of the serpent power," Marian declared. "But it passes on to the bearer of the Light from century to century. Bertie came by it in a future century, and has returned to a place in our century, where the belt lies buried."

"Can it be in two places at once?" inquired Bertie.

"More than that. In every time, potentially in every place, if there is one willing to be a bearer of the Light."

"What does it do for the bearer?" asked Bertie.

"When closed, with the silver tail in the mouth of the golden head, a circuit is closed, and a vortex of power is created through which the Light is channeled," said Marian.

"Do you know why you are here, boy?" questioned Hugh of Hawton.

"Only that Robin was my idol," replied Bertie.

"Beware of false idols," laughed Hugh, nudging Robin, who laughed as well.

"No, Robin has given him little to idolize since coming our way," said Marian, acidly.

"Well, he does become famous," Bertie stated, defending Robin.

"For what?!" demanded Marian.

"There's a book called *Robin Hood, the Great Outlaw* in the window of a bookshop in London in my time. You poach the King's deer and rob from the rich to give to the poor!" exclaimed Bertie.

"Yes, robbery really suits you, Robin," said Marian, tartly.

"But you could be so much more than an outlaw," declared Bertie.

"What else could I be?" said Robin, indifferently.

"A leader of men," responded Bertie. "You could do something for all the people in Shirewood instead of just hunting and robbing."

"You waste your breath on the idle," said Marian. "He's a *puer*, and will remain a *puer*."

"She knows I don't have the French," said Robin, "and so uses it to make me feel inferior.

"That was Latin," Marian responded haughtily.

"Enough!" said Hugh brusquely. "While lovers quarrel the Shire-reeve prepares to loose his quarrels on the people of Shirewood. Do you have the new bowstrings which will defend us?" Hugh asked Marian.

"I do," she replied, stepping into the entrance to her tree house and returning with the crystalline bowstrings. She carried them in a deerskin bag, and two pairs of deerskin gloves were in her other hand.

"Hugh. Robin." She said. "Whoever touches these, whether to string a bow or loose an arrow, remember that you must not touch the string. Wear the gloves, or you will gradually sicken and all your life's energy will be drawn out of you. But a shaft fitted to these strings cannot miss its target, for death is its legacy."

Robin and Hugh then prepared to take their leave, but Hugh had a favor to ask of Bertie. He had taken a shine to Bongo, and asked Bertie's leave to bring Bongo with them while they made new bows for the crystalline strings. Bertie had only been separated from Bongo once before, by the machinations of Captain Hook, but Bongo seemed so intent on accompanying them, that Bertie agreed.

"Oh, why don't I go too?" said Bertie.

"We have a riddle to unravel," replied Marian, "and I may need you for that."

"Well, goodbye for now, then," said Bertie. "Remember Bongo likes his head scratched from time to time. Oh! I almost forgot. I'd like to meet Little John if you could bring him back with you?"

Hugh smiled strangely.

Robin said, "Don't tell me Little John is going to be famous too?"

"Oh, yes," replied Bertie. "He's in the same book with you."

"In my book?" said Robin, unbelievingly.

They both shook their heads and departed.

"Little John?!" Bertie heard Robin say once more from a distance.

"Do you know how long you will be staying with us?" Marian asked Bertie.

"I don't know, and I'm not sure why I'm here," replied Bertie.

"Well, come in now," invited Marian, "and we'll direct our minds to Nilrem's riddle. He never says anything without there being a purpose to it, foolish though it may seem."

"I guess my being here has something to do with my purpose," Bertie said thoughtfully, "but I can't think why now."

While Marian turned her hand to spinning, Bertie wrote down the Fool's riddle as they remembered it. Try as they would, however, they could make no sense of it. Tiring of the task, Bertie took to questioning Marian.

"How did you and your company of women come to be part of the Shirewood band?" Bertie inquired.

"I am a Norman noblewoman," replied Marian, "but when my father was killed unjustly and his castle expropriated by another Norman baron, I sought the protection of the forest rather than that of the man who slew my father. My company of nine is sworn to me in fealty."

"How did you come to know Queen Eleanor?"

"Oh, Bertram, so many questions—I'll never finish my spinning." She stopped her work, and turned on her stool to face him. "I am from the Aquitaine. Queen Eleanor's father and my father were good friends, and we played together as children. She is more intelligent than any woman in England, or in France for that matter."

"Then how did you come to be in England?" persisted Bertie.

"You see when England was conquered by King William one hundred years ago, English land was given to the French nobility by King William."

"Then he's just like Hitler!" exclaimed Bertie.

"Who is Hitler?" asked Marian.

"Hitler is the man trying to conquer England in my time. God help us if he ever becomes known in England as Adolph the Conqueror. Then we'd be slaves the way the English are now. But power over people is evil. I know it because Mrs. Bedonebyasyoudid taught me that. It starts with holding someone as Grimes did Tom when he wanted to wash in the limestone spring. Then Captain Hook wouldn't let me out of the story and wanted to keep me as his cabin boy. It starts like that, but then it can become making other people slaves."

"Oh, how you go on," said Marian, amused at Bertie's outburst. "Most of the English Saxons are not slaves. Some are villeins—serfs—and some of the artisans and traders are free men. And King Henry is quite different from his grandfather William the Conqueror. He has championed the Saxons."

"How?"

"You heard him telling the Archbishop why he wanted the clergy judged the same as any other men. He believes in justice for all men."

"Well, he doesn't seem to show it."

"But he does show it," replied Marian. "Do you know how they determined whether or not you were guilty before Henry? They did something horrible to you, and if the fire didn't burn you, or the water drown you, then you were not guilty. According to that system of justice, called trial by ordeal, God would intervene and prevent your dying if you were not guilty."

"So everyone was guilty because they all died," said Bertie.

"Yes, of course, and this gave the executioners and judges a clear conscience. But King Henry uses a jury of people to listen to the evidence and then make a judgment. It's much fairer."

"We still have the jury in my time," replied Bertie, "so I guess he is a good king—but not much help to us now," he added darkly. "If I die, I won't be able to carry out my purpose. There must be something I can do."

"What is the difference between this time now and your time," said Marian matter-of-factly, "a turn of Hecate's wheel, eight spokes in her wheel, eight hundred years."

"Who is Hecate?" asked Bertie, bewildered.

"The moon goddess of death," replied Marian. "It is she who spins the crystalline bow strings for me."

"Then these bowstrings come from the moon goddess?" said Bertie.

"Yes, whatever arrow they launch flies straight to the target; they cannot miss. That is why any archer who draws the strings must wear a glove, lest he perish from Hecate's touch."

Just then a strong knock resounded from the trunk of her tree, and Hugh of Hawton stood outside with a dwarf at his side, and Bongo perched on his shoulder.

"This is Little John," he said.

Bertie was shocked.

"No, Little John is a great big man, bigger than Robin."

"I am the only one in Shirewood known as Little John," said the dwarf.

"But there must be some mistake," cried Bertie.

"No mistake," replied the dwarf.

"Why would anyone call a big man Little John?" inquired Hugh.

"Oh, never mind," said Bertie peevishly.

"I am the bearer of bad tidings, Lady," said Hugh to Marian.

"Has something happened to Robin?" she asked earnestly.

"No, Robin is well and has undertaken a mission on your behalf," Hugh replied.

"On my behalf?!" said the astounded Marian. "I asked him not!"

"If you will hear me out, Lady," said Hugh.

Marian nodded.

"Your conversation with the King has drawn the attention of the Shire-reeve. He entered the forest with a company of the Earl's knights not an hour ago. They sought you and took a man and his wife in hostage. Your safety is in jeopardy. Robin has gone by horseback to organize the length and breadth of the Shirewood in your defense. If the Shire-reeve returns, he will have a warm welcome."

"Robin! I knew he'd come through!" shouted Bertie.

"But that is not the worst of my news, Lady," Hugh continued. "We fashioned the bows for the strings you gave us, but no one in all the Shirewood is strong enough to draw those strings. They are worthless, and the Shire-reeve may return at any time."

"Not worthless!" exclaimed Marian, her face contorting from its dark side in an angry scowl. "The men of the Shirewood are worthless!"

And she turned and stalked into the tree house.

"We shall be in the trees keeping watch with our longbows this night," said Hugh to Bertie. "But you and the lady have the least to fear from the night. I do not think the Shire-reeve will be foolish enough to come in bearing torches. He and his men would be picked off like fireflies."

Bongo hopped from Little John's shoulder to Bertie's. They were the same size. As Hugh and Little John took leave Hugh turned to Bertie and said, "Little John is a woodcarver of some skill. He has an idea in mind for your monkey friend. On the morrow when Robin returns, we shall surprise you, perhaps, with Little John's ingenuity."

Bongo nodded vigorously his enthusiasm for Little John's idea, and waved goodbye to them as they went.

"What are you up to now?" Bertie said suspiciously to Bongo.

That night, fearful of discovery, Marian, Bertie and Bongo huddled in the dark of the tree house without candlelight.

"A full night's spinning wasted," lamented Marian, "and the full moon will not come again for another month. Ah, but what does it matter if no man can pull the crystalline bowstrings."

"Why do you spin at the full moon?" asked Bertie.

"That is when Hecate helps to draw her treasures from the earth," replied Marian.

"Is it dangerous?" Bertie questioned.

"There is never any creation without sacrifice, and Hecate never gives up death without taking back unto her a proportionate amount of life. That is why I wear the golden thimble to protect me from the prick of death."

"What death has she given up?" persisted Bertie.

"The crystalline bowstrings. Talismans of death that speed an arrow unerringly to its goal."

"But what good are they if no one can pull them?" said Bertie.

"No good, or we must find another way," said Marian. "I wish Nilrem were here now so that I could question him about his riddle."

"But he wouldn't remember it, would he?" asked Bertie.

"No, he wouldn't," said Marian. "Our lives are hanging by a thread—or a bowstring."

"There must be something I can do," Bertie declared. "There are so many modern weapons that could help us, but I don't know how to make them."

"What kind of weapons?" Marian wondered.

"The worst are the buzz bombs," replied Bertie. "One of them killed my mother and father."

"Do they hum like bees?" asked Marian.

"No," replied Bertie. "They fly and then fall silent just before they kill."

"They are like birds then," declared Marian. "How do they fly to where they are to kill?"

"They put wings on the bomb, and set it on a metal track pointing in the direction it is to fly in. The elevation is controlled by how high they point the nose on take-off, and also by a control that automatically makes adjustments during flight."

"And what makes it fly?" inquired Marian.

"Fuel that burns during flight. We don't have fuel and we don't have explosives," said Bertie. "I could invent gunpowder so that we could shoot bullets at the Shire-reeve's men, but finding the minerals would take years."

"And by then we'd surely be dead," added Marian with certainty.

For a long time neither said anything. It was as if the gloom of the night and their little room had penetrated into their very souls. Then Marian spoke.

"Let us recite together like a prayer Nilrem's riddle. The answer lies there, I know."

And together they recited as follows:

"To the supple woman,
Marry the upright man.
Temper the fiery metal
With the balmy yew.
Resolve the pairs of opposites,
And so shall have you,
Death in life, life in death,
Tempered to try their mettle."

"Now," said Marian, "I shall ask you questions, and you shall ask me questions about the riddle, and if necessary we shall sit here until the sun rises."

"Or until we see the light," said Bertie laughing

"What are the pairs of opposites in the riddle?" Marian asked Bertie.

"Life and death?"

"Yes, and also?"

"I don't know," answered Bertie.

"A man and a woman," said Marian.

"That's right."

"And wood and metal," added Marian. "But how is the metal 'tempered' by wood?"

Bertie thought a long time and then said, "Remember when I wrote down Robin's poem to you, and I didn't know yew was a wood? You told me it was used to make bows and arrows. Maybe this riddle is an instruction as to how to make a bow that can pull the crystalline bowstrings."

"Yes, of course," replied Marian. "They are the key to the weapon."

"What is there in your world, Marian, that can send something flying furthest of all?"

"The catapult," she replied. "It is used to send rocks flying over castle walls, or to break them down. What about something like the track which the flying bombs fly on, only it launches the arrows?"

"Like a catapult!" exclaimed Bertie.

"Yes," said Marian.

"Eureka!" cried Bertie. "The crossbow. The track is made of metal."

"That is the upright man!" said Marian.

"And the bow of yew—"

"—is the supple woman!" concluded Marian.

"Have you ever heard of a bow that is held horizontally rather than vertically like the longbow?" asked Bertie.

"No," replied Marian.

"Then we are about to invent it," said Bertie. "How stupid of me. The crossbow follows the longbow and comes before the gun in the history of weaponry. We shall need some metal parts for it. Is there a metal worker in the Shirewood?"

"Yes, the Smith, a freeman who joined the band on his own. He made the sword for Billy Budd that slew the Black Knight."

"Then it can be done," concluded Bertie.

"At first light, we shall rise, breakfast, and then search out the Smith," said Marian. "If we live to see the dawn, the Shire-reeve may be thwarted in his purpose. Do you say prayers before sleeping?"

"Sometimes," said Bertie.

"Then this might be a good time to say them," said Marian.

Bertie lay down, folded Bongo in his arms, and said his prayers.

"Now I lay me down to sleep,
I pray the Lord my soul to keep,
If I die before I wake,
I pray the Lord my soul to take."

And then he slept.

He was half-asleep when he began to hear Marian moving about, making breakfast, and preparing what they would need on their journey. When he finally opened one eye, he saw that the sun had already risen.

When they came outside, there was Hugh Hawton and Little John awaiting them, but no sign of Robin.

Hugh explained. "The Shire-reeve had spies in the wood last night, men dressed like the people, who followed Robin hoping he would lead them to you. So it was best that he stayed away."

"Yes," agreed Marian, "but is he all right?"

"No harm came to him. He awaits us at a hiding place where many have gathered to defend you," said Hugh.

"Our future is brighter today, and that of all the people of Shirewood, thanks to a fool and a genius. But we must find the Smith as soon as possible."

"Then we shall escort you to him," replied Hugh.

The Smith's forge was dug into the roots of a great oak. From Nottingham he had brought with him by wagon his tools and molds when he had joined the band. With a woven mat of dried stalks, he could cover his entrance, rendering it no longer visible to mortal eye. Normally he labored only on the windiest days, when the smoke which would betray him was blown away.

Hugh and Little John set out to find a sturdy branch of yew from which to make the bow and the stock on which the metal track would lie. Bertie drew a diagram in the dust and explained the concept of the crossbow to the Smith.

"What is it called, then?" inquired the Smith.

"It hasn't been invented until now," replied Bertie, "but it is called a crossbow."

"It will crucify your enemies, that is certain," said the Smith certainly.

"There is one difficulty," spoke Marian, removing with glove the crystalline bowstring from its deerskin bag. "No man has been able to draw this string." She held it up to the light. The Smith reached out as if to take it. Quickly Marian drew it back. "You must never touch it with your bare hands. Death will be the result."

The Smith peered intently at it. "It seems to be made of metal," he concluded. "How is it drawn from the rock?"

"In a cave at the full moon, and spun about a spindle. More I cannot speak," said Marian.

"Have you seen the catapults used to fling rocks over castle walls?" Bertie asked him.

The Smith nodded. "I helped build the one the King's men used to besiege the castle at Nottingham. The Norman lord has made my life miserable ever since, and that is why I am living in the Shirewood now."

"Well, on top of the straight stock," said Bertie, "you must lay a miniature catapult of iron that can be moved back mechanically along notches until the bowstring can be drawn tight."

"What holds it then?" asked the Smith.

"A lock in the last notch. Then you need a trigger under the stock that will release the bowstring at the mere pull of a finger once the crossbow has been aimed from the shoulder," said Bertie.

"What about your arrows?" inquired the Smith. "With a bow of such force an arrow could travel four hundred yards, only to bounce off mailed armor harmlessly. Let's put pointed metal tips that may seek out the flesh in the rings of armor as a dart homes in on the bull's eye."

The Smith's forge lay but a few hundred yards from a clearing where the people of Shirewood had grown thickets and hedges high enough to thwart the leaping deer, and closely intertwined at the base to keep out marauding rabbits Within was a vast grain and vegetable garden which augmented the diet of deer and rabbit which sustained the Shirewood's people. Wooden buckets attached to a turning wheel ladled water from a tributary of the Trent onto the edge of the garden. From there the water ran slightly downhill irrigating the entire garden through a system of shallow ditches.

However beautiful and functional, the garden was illegal because the protective hedge wall did not permit a horserider to pass through in pursuit of game. Since the entire Shirewood was the King's own forest, the Shire-reeve had made numerous forays into the wood in order to destroy the offending thickets. However, guards with longbows posted in the branches of the tall oaks overlooking the site had discouraged further attacks by the Shire-reeve. Furthermore, constantly on the move from shire to shire bringing law and order to his country, King Henry had little love and time for sports.

It was to this site that Marian, Bertie and Bongo, and Hugh and Little John now came while waiting for the Smith to make history's first crossbow. Despite the fact that knights or the Shire-reeve's men might ride in at any moment, a kind of carnival atmosphere prevailed. Vendors of rabbit, venison, and fruit pies were everywhere, and jugglers, clowns, musicians, and fools entertained. Ahead a crowd had gathered around a man perched in the lower limb of a great oak. Marian and her little party were wandering that way when Bertie exclaimed, "It's Robin!"

Haranguing the gathered multitude, he was more eloquent than Marian had ever heard him before. His argument was that the people of Shirewood should form a committee to petition the King for a meeting and hearing on their grievances with the Shire-reeve. Since they were all outlaws in the eyes of the Shire-reeve, and could be arrested on sight, they could not avail themselves of the shire courts set up by the King himself in order to mete out justice. Several hot-heads in the crowd disagreed and were for all-out war until one side or the other was destroyed.

"That way will not work for us," Robin declared when these men had had their say. "The Shire-reeve will paint our just cause with the brush of open rebellion! Then we can expect not the King's justice, but his great wrath!"

"Here, here!" agreed many in the crowd.

Bertie felt his chest swell with pride for the new Robin, and turned to see Marian's eyes bright with admiration.

When Robin had finished speaking, the crowd voiced its approval for a committee to set down their grievances and petition the King for a hearing. Dismounting from his tree perch as agilely as a sparrow, Robin made his way towards Marian, followed by a complement of men who had agreed to be her personal bodyguard even unto death. Flattered but somewhat dismayed by all the attention, Marian was quick to change the subject from herself as the center of attention to Bertie.

"Robin!" she cried, "the boy's invented a new bow that can draw the crystalline bowstrings. The Smith's making it now."

Robin looked down at Bertie at her side and said, "Is it so?"

Yes, nodded Bertie and Robin reached down and lifted him up to his shoulder, where he sat surveying the crowd like a king at coronation. Never before in all of his adventures had he felt such pride and excitement. Bongo, in his own excitement, with his arms around Bertie's neck, began to drum on Bertie's shoulders with his paws.

"Tell me about it, then," said Robin to Bertie.

"It's like a small catapult with the bow in front," replied Robin.

Robin nodded, then whispered, "But I still say the earth is flat."

Bertie laughed, not caring anymore, for Robin was truly his hero now.

Hugh of Hawton sidled up to Robin, strolling side-by-side with Marian, and as if it were pre-arranged, Bongo leaped onto Hugh's back, who said to Robin, "Yonder, by the great larch, Little John has a small stall that will interest you and the boy if you care to wander by." Then Hugh walked off.

"I wonder what that little monkey is up to now?" said Bertie to Robin and Marian.

"Shall we see?" asked Robin, but before they could move on, a pie-maker, with whom Robin had once shared a great stag he had brought down, gave them six apple tarts, topped with dripping honey fresh from the hive.

Robin got some honey on his chin, and Marian tenderly wiped it away with her handkerchief while gazing intently into his eyes. Then she looked at Bertie on his shoulder and smiled. Now she seemed to be all light suffused with an inner glow, despite her dark half of coloring, which sometimes caused the people to stare.

They made their way to Little John's stall at the edge of the crowd. An expert woodcarver, he had made a replica of the bareknuckle boxing stalls that were always the center of attraction at any country fair. Standing on a platform alongside and behind the booth, Little John could pull open the curtains which concealed the mystery within. A boxing champion was selected by the crowd, and wagers made on whether or not the mystery champion within could deliver a punch to the chin. When the curtain parted, there stood Bongo, both paws raised in a boxing pose. The crowd roared with laughter, but Little John continued to wager on his tiny champion, so still more bets were placed on the crowd's champion, who under the rules could not hit Bongo, but must prevent Bongo from delivering his "sneak punch," as Little John called it.

Little John would accept no wager more than a penny; yet eventually practically everyone in the crowd had bet on the crowd's champion so that a huge pile of pennies rose alongside the stall. When all wagers had been placed, and the time came for the match, Little John instructed the champion to bring his face up to the booth where "the boxing monkey" could reach him. "But cover yourself at all times, sir. Never drop your guard, for monkeys are the most devious of characters."

Then a bell was rung and Bongo thrust out lefts and rights, which were easily thwarted by the huge parrying fists of the champion. Then a bell was rung for the end of the round. Bongo sat down on a small stool, whereupon Little John fanned his face with a miniature towel as the crowd roared in laughter.

The next round was more of the same, a furious flurry of punches from Bongo, easily blocked by the champion, who stood stupidly grinning, except that now Bongo's gyrations had caused his boxer trunks to fall, sending the crowd into higher hysterics as Bongo reached down to pull them up.

At the end of the second round, as Bongo sat down on his stool, he was given a miniature bottle to drink from and soaked with a tiny sponge. Then the bell sounded for the third round.

This time Bongo's tactics were different. His punches were slow and deliberate as he moved ever closer to the champion's face. Finally, from behind himself, his tail to which Little John had surreptitiously attached a boxing glove, snaked out and landed on the man's nose. The crowd let out a roar of approval as the dumbfounded champion stood grinning sheepishly. Little John raised Bongo's hand in triumph, and Bongo bowed this way and that to the crowd's applause. Then the curtain closed on Bongo's career. Little John thanked the crowd for their enthusiasm and told them that their pennies would be donated to the Virgin, which made them feel twice as good as before; thus everyone departed in a happy mood.

Robin, Marian, and Bertie proceeded in the direction of the Smith's forge, trailed by her bodyguards, and moments later Hugh strode alongside to return Bongo to Bertie. Bongo now rode on Robin's other shoulder, and whenever anyone recognized or spoke to him, he gave the clasped-palms salute of a boxing champion.

"What a ham!" said Bertie.

As they approached the Smith's forge, they noticed a number of men gathered near the Smith, all of them busy at tasks which he had assigned them. The Smith greeted them, then noticing Bertie on Robin's shoulder, he said, "If the boy brings this off, we'll have to get a royal chariot for him to ride around in."

Bertie inquired of the Smith's progress, and he replied impatiently, "It's one thing to imagine a new bow, yet another to make it. Once the design is determined, molds have to be made for the metal. That way others can be made more rapidly than if

215

each had to be handmade."

Bertie thought, "The Smith has himself invented something—mass production."

"How long then will it take before you have the first new bow?" inquired Robin.

"Tomorrow afternoon at the earliest," replied the Smith. "Unless the Shire-reeve rides in and gets his hands on it first."

"That would be terrible," said Bertie. "That weapon could change history to evil's advantage."

"Then don't worry about me," said Marian earnestly. "Put my bodyguard to protecting the forge."

"If I have two to protect, then I shall put them in one place," said Robin forcefully. "You must stay near the forge this night, otherwise our forces are split in two."

Eventually it was determined that the safest place for Marian, Bertie, and Bongo was up a tree. Bongo liked the idea immensely, demonstrating to Bertie that it was possible to sleep while hanging by the tail. Bertie wasn't particularly keen on the idea, but wanted to please Robin. Marian would have none of it, and began arguing with Robin.

"Is it because you still fancy yourself a Norman lady that you feel sleeping in a tree is beneath you?" said Robin.

"No," replied Marian, "I just am afraid of falling out of the tree."

"Then we'll make you secure by binding your blankets with cords," replied Robin.

"I don't want to be bound by cords!" Marian answered angrily.

"Don't you realize the Shire-reeve's men will never climb up trees looking for you?!" responded Robin.

Eventually the nine women of Marian's company were also sent for, since they might be taken as hostages for her, and they convinced Marian to follow Robin's advice when they agreed to share the same tree with her. Men of her bodyguard filled the surrounding trees. Before she fell asleep, she thought to herself how much more pleasant were the leafy branches of a tree, how light and airy, than the close dirt floor of the tree's roots.

Far too excited to fall asleep at once, Bertie spent an hour or two talking to Bongo, who was playing Bongo Python by dropping onto Bertie from the limb overhead. Finally Bertie caught him and whispered to him about their adventure with Robin Hood.

"Bongie," he said, "we've only been here three days and already we've had dinner with a king and queen and three princes. And I've invented the crossbow and you've become a champion boxer. Tomorrow we'll show Robin how to shoot it, and tonight we have a whole tree to sleep in, and we could never sleep in a tree if we had stayed in London, never, never, in a million years...."

And then he slept.

That night Merlin came to him again. He sat on a limb across from Bertie, dangling bare feet in the air. Bertie sat up, while his reclining self continued to sleep.

Bertie thought, "Where am I, Merlin?"

"You are in England in 1163," came Merlin's thought. "The past is contemporary with the present, but do not think that you are in 1163 as it happened before. Travel is distance in space; time the duration taken to the place. Therefore, no one can return to a completed place in history. However, you are in 1163 with a real flesh-and-blood Robin Hood. You are recollecting 1163, or put another way, re-*collecting* 1163. The process is not one of travel in space, but a psychic act of high consciousness occurring on the mental plane, but in fact the levels of consciousness are deeper than the ocean, and as infinite as space.

"Think of your consciousness as the joining point of the two cones of an hourglass, where the grains of sand from the top cone flow into the bottom, when future becomes present in that moment, and then past. The light cones, everywhere at once in time, wedded to the material particles of stones, trees, castles, knights, and Robin Hood himself, are being *attracted* to the light of your consciousness, which is inseparable from the All-Encompassing Light of the Infinite Universe to which you give the name God. The attraction creates a wave phenomenon, as you call it in physics, whereby *dead* history is resurrected and lives again.

"But by viewing history, or simply living it as you are now, your consciousness—inseparable from Light's Creator Consciousness—creates the events of *your* life. The hourglass of your mind funnels in the particles of the so-called past, recollecting it and re-*creating* it through the *now* moment of your consciousness. In the now moment, it cannot be the same past as before, because—for one thing—you were not there before in 1163. And even if you spoke to no one, and no one saw you, your footprint would still be in Shirewood, and your shadow would fall down through history upon every century thereafter."

Bertie thought: "It is an awesome thing that I have undertaken."

"Indeed, it is," came Merlin's thought. "But your Purpose has prepared you for it, and beyond even the Purpose known to you now, the Light within knows that to prepare you for bringing new models of space/time to the world, your education—shall we say—must be unique, indeed.

"All the warp and woof of history from the King on down to the Fool, will be changed by you."

"Robin?" thought Bertie.

"Robin you may be sure, and all the band, and all their antagonists as well—even the future. The continual battle of protagonist/antagonist, Light and Darkness, is a cosmic *game*, whereby Light acquires greater consciousness of Itself. The game suits world history as well as distant galaxies where Black Holes seek to swallow Light, and so the battle rages evermore and eternally."

"Does Darkness ever win?" Bertie thought.

Merlin's thought: "What a silly question from one from whom Light expects so much. If Darkness did not win, there would be no game."

Then a dark thought flashed across Bertie's mind: "Must I kill?"

Merlin's thought came back at once. "If you are not prepared to kill, then others will surely die, often loved ones who will be the victims of those it was your moral and spiritual choice not to kill. Then it is as if *you* had murdered those loved ones.

"The choice then is to oppose—or not—the Darkness, and thereby win a victory for the Light. Lives lived without making this choice must be repeated through many lifetimes until one is at last capable of making a decision."

And then Bertie almost laughed, because his thoughts had awakened Bongo, who sat up and looked at Merlin. Bongo's thought came to Bertie, and he asked it of Merlin:

"Bongo wants to know if he will ever become a boy?"

Merlin's thought: "Not in your lifetime with him, but ultimately the answer is yes. Because of the gap in space/time that you are in, your inanimate companion is acquiring consciousness which he is utilizing with a very strong will to evolve into a higher form. Sometimes a very unique consciousness jumps over the next biological step in evolution, in a kind of cosmic leap-frog, or like skipping a grade in school.

"Light is the first quality of everything in the universe. It imbues every cell with its quality, from the lowest evolutionary creature to the highest. It precedes matter and space, and even time. Light is the Creator principle. No creature, no particle, is separate from the Creator because it is endowed with the quality of Light. Time and space are really qualities and not quantities because they are qualified by anyone's reaction with them. Remember that when you begin to formulate your theories of space/time when you are an adult."

Bongo nodded in agreement, and lay back down in his sleeping form.

Bertie's thought "What does it mean that I am in all these books?"

Merlin's thought/reply: "You are the author of your life. Goodnight."

And Bertie lay down in his sleeping body and remembered nothing until morning when he asked, "Bongie, was Merlin talking to us last night?"

Now today if you journey to what remains of the Shirewood, now called Sherwood, and are fortunate or unlucky enough—as the case may be—to sleep in the wood, at dawn you may see that same sight that Bertie saw in 1163. The bark and leaves of the trees take on a warm glow like flesh; then the limbs begin to stir like sleeping men awakening after many centuries. Some say it is only the wind; others that is the men of Robin Hood's band bestirring themselves after the long night's sleep. That morning, after their initiatory night in the tree, Bertie and Bongo were truly of the band of Robin Hood.

Bongo Python had awakened before Bertie and was waiting for him to open one eye before he sprang down upon him.

"Ooof!" said Bertie. "Get off!"

Down on the ground they met Marian, and Robin, who had slept at the base of the tree. Surprisingly she was smiling.

"It's not so bad, is it?" she said to Bertie.

"Bongo kept jumping on me playing python," complained Bertie.

Marian and Robin laughed. Then down came the nine women of Marian's company, laughing and bantering with one another. All eyes turned upon them, for each of them was fair indeed. Breakfast of pork sausage inside bread rolls was welcomed by all except Marian and the nine, who said that the pig was sacred to the Mother Goddess and they would not eat the flesh. Instead they gathered berries while Marian made tea.

Despite trying to preserve an attitude of calmness, Robin was just itching to call in at the forge to check on the Smith's progress. Finally he could stand it no more and called Marian and Bertie to him.

"Shall we see how the Smith is progressing?" he said.

When they arrived at the forge, the Smith looked red-eyed and haggard.

"It would have taken longer than we anticipated," he said wearily, "so I have worked through the night. The fiery iron has been poured into the mold, and when it has cooled, we shall marry it to the wood."

Bertie looked at Marian, and she returned his glance knowingly, for the Fool had used the word "marry" in his riddle.

When the "upright man" had been removed from his mold and cooled, the Smith laid it in place on top of the stock, which had been perfectly carved with grooves into which the metal track fitted exactly. The track was further secured to the stock by wood, and then the bow added at the front. Finally it was ready for stringing.

"Let me do it," said Marian, donning the deerskin gloves. "I brought forth these crystalline strings from the cave, so it is only appropriate that I fit them to the bow."

When that task was completed, the last delicate operation was the adjustment of the trigger that released the bowstrings. Then they were ready to try it.

"We should set up a target," said the Smith. "Then by trial and error I can make you a sight so that it will not be hit or miss."

"It doesn't need a sight," said Marian.

"Why not?" questioned the Smith.

"Because the bowstrings will send the arrow flying to the man at whom the bow is pointed."

The Smith gave her a strange look, but did not contest the point further.

"You won't need to try it first," said Marian to Robin. "When the time comes, death will be in your service."

Robin nodded.

"Then," said the Smith, "I bid you *adieu*, for I may sleep until the next sun-up."

But before he could retire, congratulations were offered him and his woodcarvers by all in attendance.

"I must know your name," said Bertie to him, "so that it can be placed in the history books."

"Weyland," said the Smith, modestly.

At midday Robin convened a meeting of those men desirous of serving on the committee to present the King with the grievances of the Shirewood's people. Marian protested.

"Do not the women of Shirewood have grievances too? Therefore, there should be at least one woman on the committee," she said.

"But none care to serve," replied Robin.

"I do," responded Marian. "And I have already bent the King's ear, and have a channel to him through the Queen."

It was agreed then that Marian should serve on the committee, and later when each man was through speaking, she brought to their attention that each man who had been involved in conflict with the men of the Shire-reeve was now a marked man.

"In future, I suggest," said Marian, "that masks be worn by the men who oppose the Shire-reeve's men. Then they cannot be marked and hunted down like wild animals. Instead we all share the blame, and they must deal with us all."

"And where shall we find masks in the Shirewood?" one man asked facetiously.

"My company of women shall fashion them," replied Marian.

Robin wished to spend the afternoon testing the distance and accuracy of the crossbow, but he soon found that the metal-tipped quarrels buried themselves so deeply into his target trees that they were wasted because they could not be pulled out. He then sent out men, women, and children in all directions to bring out yew wood suitable for making arrows. Near sundown the Smith awakened and fired-up his forge in order to fashion throughout the night the metal tips for the arrows. All around the forge, seated on the ground and in the trees, men and women made arrows from the raw yew wood that was steadily coming in from all over Shirewood. Because the crossbow took a much shorter quarrel than the longbow, three times as many arrows could be made from the same amount of wood. Affixing feathers that would steady the quarrel in flight was the work of the older children, and Bertie was helping with this task when Marian came up and invited him to accompany her. Bongo had already gone off with Little John to box again, so Bertie readily accepted Marian's invitation.

Later, he had second thoughts when he learned that she was going in search of wasp nests.

"They sting!" he told her, as if she did not know it.

Then they were joined by the nine women of her company, all carrying bags of plants, leaves, and roots, which stained the bags with their bright colors. Walking along a dry riverbed, they kept an eye out for nests hanging from overhead limbs. Because the trees did not reach out to the center of the riverbed, they could see along the limbs many yards ahead of where they walked. Presently two nests were spied almost opposite one another in two giant oaks.

Going to the first tree, one woman began climbing while the nine others joined hands in a circle around the tree. As Bertie watched, he heard a low humming coming from the circle of women. As the humming grew louder, the wasps began to fly from the nest, and by the time the climber had reached it, she was able to carry it down, where it was placed in one of their bags. The same procedure was repeated at the next tree, and so on, until they had filled many bags with nests.

By nightfall they had returned to the camp, where Little John awaited Bertie with news of Bongo's triumph against three challengers.

An uneventful night was spent in the tree, and when morning dawned, no sooner had Marian and the company of nine completed breakfast than they set to work fashioning the masks that would disguise the wearer's identity from the Shire-reeve's men.

There were nine nests in all, and nine masks when they were finished. Bertie had wandered over to the forge after breakfast to find that Robin now had over one hundred arrows at his disposal.

"The problem, you will find, Robin," said Weyland the Smith, "is that in the time it takes you to cock the crossbow you can shoot three arrows by longbow. We need more crossbows, but there is not refined metal enough."

"I shall take these quarrels up strategic trees that overlook the fields on the edge of the wood where the Shire-reeve is known to come in. Then I can begin knocking down men from four hundred yards out," said Robin.

"Ay, that will learn them to keep a safe distance from the wood," laughed the Smith.

"Come on Bertie," said Robin, "you and Bongo can help me stow the quarrels now that you know how to climb trees."

They followed the path skirting the thicket fence surrounding the vegetable garden until they came to the edge of the forest some five miles away. There Robin indicated a huge oak with a long, nearly horizontal branch halfway up. Each of them carried a bag of arrows around his waist as they started up the tree. Bongo couldn't help showing off, climbing to the top of the tree while they were still not halfway up.

"Here, leave the arrows on the big branch and go down and get some more," called Robin to Bongo.

Bongo scurried down for a second bag. Dumping the contents of each bag into one, Robin hung it on a knob of the limb, out of sight from below. They were straddling the limb, inching along outwardly so that Robin could explain his strategy to Bertie. Because the surrounding fields fell away from the wood, one could see for several miles.

"Now, if I begin to fire off my quarrels from here," said Robin, "and men begin falling, the Shire-reeve will think that there are men hidden in the fields who are firing at him, because the forest is too far away for an archer firing from there to reach him."

"Suppose they spot you," said Bertie.

221

"They could only tell where I was by the downward flight of the arrow, if I were foolish enough to shoot while they passed under the tree. When we have more crossbows, we must keep a man in this tree."

They were about to descend, when Robin said, "Look at that!"

There, on the outermost tip of their limb, Bongo was standing on his head, his paws gripping the limb.

"He's really showing-off a lot lately, and he was so well-behaved in *The Water-Babies*, better even than Tom and I."

"Perhaps it's getting back in the trees that has done it to him," speculated Robin.

When they had returned to camp, Marian and her maids had completed the masks, which they tried on for Robin and Bertie. A pliant vine attached to the mask by beeswax held it in place on the wearer's head. Juices from collected roots and plants provided the needed colors. Horns and snouts had been stuck on by beeswax.

When the women all donned the masks at the same time and looked up at Robin and Bertie, an uneasy feeling went through Bertie, and the longer they looked at him the more panic began to rise in him. He wanted to run. Then they took off the masks. Bertie looked at Robin, and Robin at Bertie and they knew that they had shared the same terror.

Each mask was of a different animal: wolf, stag, badger, *et cetera*, but rendered in such a way as to promote panic and terror. When they were not worn, the women kept them bagged, which was good considering their numinous power. Recognizing a force unknown to him, Robin wisely placed Marian in charge of the dispensation and use of the masks.

That very next day they were to be needed. Shortly after breakfast Robin had climbed into the huge oak where he had left arrows the day before. Bertie and Bongo stayed behind with Marian and her maids, and the bodyguard assigned by Robin. Besides his crossbow, Robin had with him a deerskin waterbag and food to last two days.

Presently, in the distance, he saw horsemen riding abreast, and then something that chilled him. Glinting off armor in the early morning light, the sun betrayed the presence of knights, bearing the banner of the Earl of Nottingham. With them was the Shire-reeve and some of his men.

"Soooo!" said Robin, "now he's got the Earl doing his dirty work for him."

Robin cocked the crossbow, and began sticking quarrels inside his belt. Pointing his bow at the knight on the left flank, the one carrying Nottingham's banner, he let fly. He had aimed not on a straight line, of course, because the riders were 350 yards away, but well above. Like a falcon the quarrel flew, then dropped like a bird of prey adjusting its flight to meet its prey. Robin did not see it strike, but the rider silently fell from his saddle. Recocking, Robin aimed in the same way at the rider on the right flank, who fell as silently as his companion.

Onward came the riders, unmindful that the two flank horses abreast of the pack were now riderless.

Once more Robin fired at riders on the left and right flanks, and once more they fell silently.

"Is this magic?" he thought. "A man does not die without crying out."

Four were down, still unnoticed, for the Shire-reeve and his men in the middle were riding slightly ahead of the flanking knights as if leading the way.

The knights were twelve in all, with twelve of the Shire-reeve's men. Twice more Robin fired at the flanks and twice more the riders fell, but now they were close enough for Robin to follow the flight of the quarrel. Now, more like a homing pigeon than a falcon, the quarrel had flown in the narrow visor of the knight's helm.

With his foot caught in the stirrup, the last rider could not fall to the ground. His horse stopping, the next rider turned to see what had become of his fallen companion. Knights were strewn on the ground to left and right of him like fallen chaff. Sounding the alarm, the knight whirled his horse around and cantered back in the direction in which he had come, followed by the other five knights and the Shire-reeve's men.

Robin looked down at the crossbow. A feeling of great awe overcame him. "Is it the bow or the crystal strings?" he thought.

Moments after Robin's encounter, a man on a donkey cart came riding past the thicket garden, furiously whipping the beast.

"The Shire-reeve's men and knights!" he cried.

"Where?" the people all shouted.

"Coming from the West!"

There were two roads leading out of the Shirewood from the garden thicket. Robin had stationed himself at the southeast entrance. The other, to the northwest was unguarded, for the Shire-reeve had never entered there before.

Thinking to do the King a favor, since the Shirewood was within his hunting realm, the Earl of Nottingham had sent in his knights with instructions to harm no one, but to set fire to the garden's protective hedge that prevented deer, foxes, and huntsmen from passing through. The Shire-reeve and knights whom Robin had waylaid was to have arrived at approximately the same time in order to provide the authority for the destruction of the hedge.

Marian and her company of nine were inside the hedge working in the garden with Bertie, when news of the approaching knights reached them. The bodyguard stood idly by watching them work, but they bore ordinary bows and arrows, of little use against the armored knights. Quickly Marian gathered her women to her. The bodyguard was mustered inside the northwest wall of hedge where the knights would be passing. Still others followed Marian, Bertie, and three of her company, into the trees overhanging the path.

Moments later hoofbeats were heard on the path, and twelve knights, bearing the Earl's banner rode in. The bodyguard, and whatever other men were present, drew their bows, and waited.

As they came alongside that part of the hedge where Marian waited, she gave a signal to those in the trees and to the others next to the hedge. Supported on the shoulders of the bodyguard, the women suddenly thrust their masked heads above the hedge, to peer down at the knights. At the same moment those masks in the tree loomed down from above.

The riders stopped, then turned nervously in small circles. Then panic broke out. Between the trees and the hedge, the pathway was narrow and the lead horseman tried to turn back through the pack at the same moment that those in the middle were turning in tight circles.

Then the panic spread to the horses. Their eyes bulged at the bizarre masks; they reared and whinnied, and when another horse bumped their flanks they kicked in terror. Then they bolted back up the path whence they had come

That night men from Lincoln entered the Shirewood with news of the six slain knights. Robin had watched their passage under his tree. Without horses and dressed in the Lincoln green of their shire, he knew that they were not a danger to the people of the Shirewood.

No one feared the return of the knights that evening, so a large fire was made, around which the people gathered for supper. Afterwards, when the Lincoln men had been fed, they told the tale of the fallen knights.

One, named Adam, began speaking.

"They lay on the ground as they had fallen, forsaken by their comrades. Though their visors were down, in the slit therein, directly between their eyes, was a small quarrel, no longer than my palm."

The gathered throng exhaled a cry of wonder. Bertie knew that it was the work of the crossbow, but he wondered how Robin had been able to creep close enough to shoot into the narrow opening of the visor. Marian, however, knew the answer, and smiled enigmatically.

"The Shire-reeve and his men," said a Lincoln man named Seth, "rode like mad all the way back to Nottingham. And then there came after them another group of knights who said they had seen the Devil here."

Now the crowd laughed, knowing it had been the masks of Marian and her maids that had routed the second company of knights.

"Ay, ye may well laugh now," said Dick, the third Lincoln man, "but the Earl has sent out word to the other shires surrounding Nottingham to send knights into the Shirewood to kill every man, woman, and child in revenge for the killing of his six knights."

Those gathered about the fire gasped, and the children sought the sheltering arms of their mothers.

"The Earl cannot do this!" declared Hugh of Hawton. "The King's law does not permit it. If his knights have been murdered, then the *curia regis* will hear the case. If they have been killed trying to force their way unlawfully onto the King's land, namely his hunting realm the Shirewood, then *they* are outlaws and may be opposed.

225

Only the Shire-reeve has authority here."

"Ay!" agreed the throng.

"Ay, the law is just, but tomorrow there may be none left to tell what the Earl has done," said Adam the Lincoln man.

"What more know ye of this plan?" asked Weyland the Smith.

"On all four sides," said Seth, "the Shirewood is to be besieged by knights from the earls to the north, east, south, and west. They will enter by every road, followed by foot soldiers who will comb the wood where the knights' horses cannot pass."

"The Earl has been very clever," said Dick. "He's placed a huge bounty on the head of the knights' killer, and said that the man is hiding in the Shirewood and is a Dane."

"A Dane!" echoed the crowd.

"Ay, a Dane," said Dick. "Little love to be lost on one of the race of our former conquerors."

"This is not the first time we of Lincolnshire have heard of the travail in Shirewood," said Adam. "Elsewhere in the land there is peace, thanks be to God and good King Henry. Why are you all here? Have ye not tenant farms to work, and are there not freemen among ye with their own land to work?"

"Ay!" spoke up Hugh of Hawton. "We did have land until this year when we could not pay the geld tax, and so were thrown off our land by the Shire-reeve."

"Ay!" said another man. "And now the Earl of Nottingham owns the lands we have lost."

The Lincoln men looked at each other, puzzled expressions on their faces.

"The geld tax, did ye say?" said Seth, again looking at Dick and Adam.

"Ay, the geld," replied Hugh.

"But the King cancelled the geld this year," declared Adam.

"What?" cried Weyland.

"The King's last land tax on freemen was in 1162," said Dick. "Now tenant farmers pay their earls, and the earls pay the king, and the knights pay scutage in lieu of military service for the king. But no more geld."

"Why!" cried Marian. "The Shire-reeve showed me a paper doubling the geld for this year. That's why we're here."

"Ay, that's why we're here!" went up the cry from the multitude. They were talking loudly among themselves when Marian, with the help of Hugh and Weyland shouted for order.

"There is not a moment to be lost!" she shouted. "Where is the man with the donkey cart?"

He stood at the back of the throng.

"Listen to me! You must ride out of here now and find the King!" cried Marian.

"I could not find my way out of here at night, and my donkey would break a leg in some hole or other," replied the man.

"Then you must give up your cart and donkey to one who knows the way! The

King must be told to come to our rescue or we shall all be slaughtered by tomorrow's eve!"

"I'll go!" said Hugh. "I know the Shirewood like the palm of my hand."

"All right," said Marian. "Here, take my safe-conduct badge given me by the Queen."

"No, wait Marian!" said Bertie. "It's a law of physics. The donkey cart will go farther and faster with a lighter man aboard. Little John should go instead of Hugh."

"Of course," agreed Marian. "Where are you, John?"

Two men in the crowd hoisted him up to where he could be seen by Marian.

"Do you know the wood?" she asked.

"Like Hugh of Hawton's other palm," replied John, and everyone laughed. "But where is the King, my Lady?"

"He left Lincoln this morning, my Lady," said Adam, "before news of the dead knights reached town."

"Ay," said Seth, "and he was heading towards Cambridge with the Archbishop of Canterbury."

"Then listen, John," said Hugh. "You must go to Newark-on-Trent, thence south to Gratham, and then over to the main road south from Lincoln to Cambridge. Where the Grantham road joins the main road, just south of there at Folkingham is an Augustinian abbey. The Archbishop would want to stay there. Go there, and if the King has not already left, you may yet catch him."

"But that is a day's ride by horse," said Weyland. "He'll never catch the King by donkey cart."

"As soon as the chance presents itself, John," said Marian, "you must buy or *steal* a horse."

"Ay, we'll pass the hat," said Weyland.

"He cannot ride a horse," scoffed a man in the crowd. "His feet can't reach the stirrups."

"Then hold to the saddlehorn, John!" cried Hugh.

"And pray to the Virgin mightily," laughed another man.

"I have now with me all the pence won in wagers against our champion, Bongo. I had meant it all for the Virgin, but perhaps she will forgive me if I buy a horse to take me to the Archbishop."

"Ay," laughed a man in the crowd. "Attain the Archbishop tomorrow, the Virgin the day after, and thus by a slow progress, heaven thereafter."

"Oh, if only Robin were here. He's no help to us now," lamented Marian.

"I know where he is," said Bertie. "I took the arrows there with him yesterday."

"Then tell John," uttered Marian.

"But is it on his way?" inquired Hugh.

"Straight out this road," said Bertie.

"That's the way I'm going," affirmed John.

"All right, where the forest ends is a big oak overhanging the road. Robin is up there," said Bertie to Little John.

"Call to him, and tell him what's happened, and to come back as soon as he can," said Marian to John.

Within two hours after John had left, Robin was back in camp. Impressed as he had been with the performance of the crossbow, he was more impressed with Marian's role in the routing of the knights by masks.

"But why," he said to her, "would six quarrels fly right in the visors?"

Marian smiled enigmatically.

No one in camp slept very well that night, less from fear of a nocturnal assault than from the excitement at the thought of the next day. After breakfast Robin presented a plan for their defense. When the Idle River had taken a new course many years ago, it had left a dry bed surrounding what had once been a small island, thereby making a natural rampart ten feet high from the bottom of the riverbed to the top of the bank. Attacking horsemen would have to climb this bank, or dismount and lead their horses up the bank, all the while under attack from the island's defenders. Here they would make their stand.

Branches were cut and brushwork defenses erected between trees. The island's occupants were plentiful enough, and the island small, so that no stretch was left unmanned, or unwomaned for that matter. When the defenses had been finished, Robin set up an archery school for the women and children. Everyone would have to help if they were to survive.

"We don't have to go to the school," said Bertie to Bongo, "because Robin has already given us personal instruction. Do you remember splitting his arrow? But if we get out of this, I'll buy you a popsicle in Kensington Park."

Bongo nodded, indicating he would hold Bertie to his promise.

Now all the yew wood collected days before was to stand them in good stead. Willow was plentiful on the island, and used to make the women and children's bows, while the stronger yew was saved for arrows.

Preparations for their defense completed, the occupants of the island sanctuary dined on venison and wild boar. At sunset Robin ordered all fires extinguished. Fear crept into the camp with the lengthening shadows. The crack of a stick on the opposite bank! Was it an animal or the enemy? The crunch of leaves! The hooves of deer or a horse bearing a knight with drawn sword? The watch was kept in pairs of women throughout the camp, so that the men would be fully rested for combat. Expecting at any moment to see knights' black helms looming in the light of the cloud-veiled moon, the women saw nothing. Morning's first light was welcomed, even if it guided their executioners to them.

Low spirits did not make good warriors, Robin knew, so he moved from station to station around the camp. To each man, woman, and child, he brought cheer and hope.

Around noon, a cry came from the watch to the east, where Robin had fired upon the knights. Down the road which Little John had traveled in search of the King, a party of knights was moving towards them, revenge in their hearts for their fallen comrades. Unknown to Robin, this was but half of a raiding party which had left Nottingham at dawn. The other half, consisting of fewer horsemen and more foot soldiers, was even now moving through the wood towards them from the south.

To the north, ten times the number of Nottingham knights—these from Yorkshire—at dawn had begun a sweep southwards, and were now at Edwinstowe but three miles away. Now this company split also, one phalanx moving west on the road to Mansfield, there to sweep east, joining with the Nottingham knights. So the trap was set, and now it was sprung.

But Robin was ready. He had placed Bertie and Bongo in the group with Marian and her company of women, who still had the terrifying animal masks. Robin had calculated rightly that an assault would be made from the road to the east, and had positioned himself there with some of his best archers, but he did not know about the knights moving towards them from all four sides.

Not wishing to betray their presence, Robin did not fire on the first knights to appear. These had patrolled the garden's hedges, poked into the Smith's forge, and finding no one, were about to move west towards Mansfield when a foot soldier, part of a company scouring the denser woods along the road where the knights could not ride, spotted them and gave an alarm which had only left his throat when Robin's quarrel found him.

Immediately Robin began firing the crossbow at the most distant knights, signalling the others not to shoot, in order to decoy the enemy into thinking the attack came from the opposite direction. Some six knights he brought down in this manner, until the last quarrel was seen in flight; then their captains moved them from out of the open into the cover of the woods.

Only minutes later they began appearing, dismounted and leading their horses, among the trees on the opposite bank of the riverbed. Now they were fit targets for Robin's longbow men, and wherever they showed themselves they fell. Now foot soldiers bearing bows were brought up, and they loosed their arrows at Robin's men, but their positions had been well-disguised by leaf and branch. Stymied for the time being, these forces did not try to cross the riverbed, but waited for reinforcements which would close the trap behind and to the sides of both defenders.

First to arrive was the rest of the Nottingham contingent from the south.

Entering the same clearing where the six fallen knights still lay, Robin's crossbow cut into their single-file column like an avenging scythe. Because of the accuracy of the arrows finding his men, the captain of this company thought them to be besieged from the trees close by at the edge of road. Thinking to outdistance his attackers, he foolishly ordered a forward charge, which brought the company within close range of Robin's longbow men.

Zing! Zing! Zing! Arrows flew thick as bees, finding horses as often as men, but once on the ground the knights were picked off one by one. The company from the south had now been decimated, and those wounded still able to ride fled back down the road whence they had come. Those knights in the wood to the east, now fanned out to the north to await the expected arrival of the Yorkshire men.

Robin watched this movement of men, and gradually swung his contingent of longbow men away from the south towards the northeast.

For an hour, nothing happened, the attackers content to shelter behind trees. They soon learned that to show themselves was to die, for Robin's quarrels found them at once. By now he had shot half a hundred of those made for his bow. If more knights came, and he ran out of quarrels, ordinary longbows would be less effective against armor.

Robin's swing of forces to the north now brought him within sight of Marian and Bertie, who had been positioned where the least action was expected.

People now began arriving from the northwest who were being swept ahead of the Yorkshire knights. And that very morning farmers in the fields around Mansfield learned of the impending assault by the Yorkshire knights, and decided to throw in their lot with the Shirewood's peoples. Robin's lookout to the west directed them to the riverbed sanctuary. Thus in one fell swoop, Robin gained two score skilled archers. But the good news of their arrival was dimmed by news of the imminent arrival of a small army from Yorkshire.

Robin consulted with Hugh of Hawton and Weyland the Smith, his two field captains. They conceded that a simultaneous charge by the Yorkshire knights from north and west, augmented from the east by what remained of Nottingham's company, would overrun them. Once the mounted knights were among them, slashing this way and that with broadsword, they would be cut to ribbons. But for now there was nothing to do but wait.

Despite their desperate situation, Robin still kept a cheerful face.

"We can expect more company from this direction," he said. "Perhaps you'd better make tea for our guests."

"Tea of wormwood," Marian replied.

"Now you remember what I taught you," he said to Bertie, while he pulled his bow. "See the arrow flying to the target before you release it."

Bertie nodded. Seeing Bongo holding his tiny bow, and recalling how he had split his arrow, Robin said, "He doesn't need any help."

Now the lookouts came running in from the north, trying to cross the riverbed. But between them and safety stood the knights and foot soldiers of Nottingham's company, and they were cut down before they could reach the riverbank.

Now the woods to both north and west were astir with the arrival of the Yorkshire army. Their captains soon learned the placement of Robin's defenses from the Nottingham captain, and strategy was plotted. Remaining out of sight, arrows were launched with a high arc, falling on many of the defenders from all three sides. The arrows came not singly, but in flights, and each time more defenders fell. Hardest hit were Robin's longbow men in the upper branches of the trees, for they had little to shield them from the rain of arrows from the sky.

While these attacks continued intermittently for the next hour, the felling of trees could be heard constantly.

"What means that?" Hugh asked Robin.

"I know not," he replied.

"It means they are building ramps," Weyland declared.

Sure enough, soon foot soldiers began scurrying down the opposite bank, carrying between them three logs lashed together and just wide enough to allow a rider to pass over it once in place against the defenders' bank. But placing the ramps was another matter. As soon as the soldiers appeared, a hail of arrows met them. If they survived this first fusillade, the nearer they approached the heavier the firepower that met them, for the children then could reach them with their tiny bows. In this manner, Bongo had deftly placed two arrows in attacking soldiers. Abandoned now, the ramps lay scattered in the riverbed.

"Have we beaten them, Marian?" Bertie asked her during the battle's lull.

"Hear that?" replied Marian, as the woods rang once more to the sound of axes and falling trees. "They will return with some new treachery."

Soon the extent of the knights' heartlessness was made known, for ahead of them, bearing the newly-cut ramps, they marched Shirewood women taken in the sweep southwards from Edwinstowe.

"Hoy!" called one. "I am Maid Martha of Edwinstowe. Don't shoot!"

"Ay, Robin, you wouldn't shoot an old woman? I'm Dame Morag."

"I'll have none of it!" cried another woman refusing to carry the ramp. "We are like the Judas sheep leading its drove to slaughter. God bless you, Robin."

"Then let your throat be the first cut," commanded the knights' captain, and she was put to the sword. In terror the others picked up their ramps and began dragging them towards the defenders' rampart. Among the defenders, men, women, and children lay down their bows, knowing full well death was imminent.

On came the hostage women. Upon the bank, the knights' chargers pawed the earth, ready to leap from the bank and up the ramps into the nearby helpless defenders. Wherever a knight showed himself, Robin's crossbow sang a litany of death, but there were too many knights, and the longbow arrows fell from the armor harmlessly as chaff. Now, within a few yards of where the hostage women would drop their ramps, there arose before them the terrifying spectacle of animal heads, and a droning singing filled the riverbed like the sound of a ghostly choir.

Shrieking, the women dropped the ramps and fled up the opposite bank, followed in turn by those knights who had seen through their visors the same ghastly visages.

When the distant shrieks had died, an ominous silence pervaded the wood.

"Have they gone for good?" Bertie asked Marian, who removed her animal mask and placed it in its protective bag, sheltering friendly eyes from its power to panic.

"I don't know," replied Marian.

Robin came to her side. "Thou hast been our salvation," he addressed her in the familiar form of speech reserved for members of the same family, or for lovers.

Marian took no offense, but smiled back, replying familiarly, "Thou has saved us too."

Bertie did not know the meaning of their idiom, but their smiles told all.

"Bongie," he said, "thou hast shot well for a dumb monkey who should never have opened a book and dropped us into it."

"While we still have the chance, we must retrieve those ramps before they can be used again," ordered Robin.

Within minutes, all the ramps had been pulled up and claimed by the Shirewood defenders. As the lull in battle wore on, confidence began to build among the defenders, soon shattered by the ominous ring of axes. Now the point of attack was being extended further and further along left and right flanks.

Most of the panicked hostages had climbed trees and eluded pursuit, but the leader of the Yorkshire knights had one more trump card to play. Of course, most of his men had not been in the front rank of attackers, and hence had not seen the animal masks which had panicked the others. Once he had regrouped his men, he ordered new hostages brought to the riverbank.

Presently a chilling sight confronted the Shirewood defenders. There on the opposite bank, standing next to freshly-cut ramps, stood the new hostages, their heads covered by black hoods within which there were no eyeholes.

At a given signal, at a dozen sites along the riverbed, stumbling and rising again, the hooded hostages, now more dreadful in appearance than the knights themselves, moved inexorably towards the defenders' rampart.

"This is the end," said Hugh of Hawton to Weyland the Smith.

"Adieu, friend," replied Weyland resignedly.

Once they had placed their ramps, the hostages came on into the defender's camp, where they were unhooded and welcomed as the friends they were.

"Forgive us," said one, making the sign of the cross.

"You had no choice," replied one of the defenders.

No sooner had the ramps been placed than the knights charged. One after the other, columns of riders leaped from the bank into the riverbed, followed by swarms of foot soldiers shooting arrows as they came. The riverbed was teeming now with armored knights like silver sharks seeking the ramps that would release them to the pool of Shirewood people awaiting slaughter. At some of the ramps, defended by Marian's masked maids, panic still ensued, but the ramps were too numerous and arrows lofted by the foot soldiers claimed many of the company of nine.

Now Bertie saw Marian bending over the edge of the bank, her golden thimble held in both hands. Climbing the bank behind her, dagger in hand, a foot soldier prepared to strike her unprotected back. Without stopping to think, Bertie raised his bow and fired. The man fell forward on his face, two arrows in his back. Bertie turned and saw Bongo, bow in hand.

Now Marian spat into her thimble, chanting over and over, "Fill, fill, fill!" Water began to trickle from the thimble, then a stream, a river, a raging torrent that coursed to the top of the riverbed's banks, engulfing the knights. Weighed down by their armor, they sank beneath the waves. Their frightened horses swam to the ramps, there to be claimed and tethered by Robin's men. As swiftly as it had appeared, the flood subsided, leaving a muddy river bottom, wherein lay the drowned knights and foot soldiers.

Then there came the clear call of a hunting horn, ringing from the south.

"Reinforcements!" cried Hugh to Robin in despair.

"We are doomed," answered Robin.

Then the remaining attackers began to turn. Behind them came on knights bearing the King's banners, who drove the attackers before them.

Then the King himself appeared, Queen Eleanor riding a white charger beside him. Behind them on a pony, waving to one and all, rode Little John.

Soon the King and Queen learned from Marian of Robin's bravery, and how arrows from the bows of Bertie and Bongo had saved her life. Then Bertie was lifted into the King's saddle, and Bongo placed before Little John on his pony, and thus ensconced they paraded in triumph with the King. Never before in history was the sight of a king more welcome to his people.

Now the Shire-reeve still did not know that the King now knew of his false collection of the geld. Therefore, he was puzzled when the King ordered him to appear before a special session of the court of the exchequer. He was made more uneasy by the convening of the *curia regis*, the King's court, for later that same day, before which the Earls of Nottingham and Yorkshire also had been ordered to appear to answer charges connected with the attack on the people of the Shirewood within the province of the King's land.

The exchequer took its name from its resemblance to the game of checkers. Squares upon a table indicated estates of land owned by the King. Counters piled upon these squares denoted monies owed to the King for the various taxes he might levy. Serfs working plots of land on the estates of earls and nobles also paid a tax for the use of that land.

In his overwhelming desire to be ever fair with all of his people, be they knights, nobility, serfs, or the lowest of villeins, King Henry was careful that no man was taxed more than he could bear. The King assumed – nay, demanded! – that collection of the monies due the Crown be accurate and honest; hence, one may well imagine his great wrath at the Shire-reeve for pocketing the geld tax, which the King had earlier rescinded.

When the day came for the Shire-reeve to appear before the exchequer, he was startled to see the King himself in attendance, and more startled still to see present many of the Shirewood's residents, including Maid Marian. Bertie and Bongo sat next to Marian.

The Shire-reeve's main duty was to collect on the King's behalf the taxes due him. With thinly-concealed anger, the King explained to the Shire-reeve that the counters before him represented monies which he had turned over to the King's agents for taxes due, "according to the records of your accounting," said the King in a rising voice.

The Shire-reeve nodded.

"Next to your counters, are placed others which indicate a great deal more money due me for the land tax on freeman's property known as the geld."

"But your majesty," protested the Shire-reeve, "you rescinded the geld tax this year."

"According to testimony which has already been heard by this court, these good people of Nottinghamshire claim you collected a geld tax from them. And when they did not have sufficient funds to meet the geld, their land was forfeited, not to the Crown as is customary, but to you personally; for you are now in fact in possession of those same lands, are you not?"

Instead of replying, the Shire-reeve shouted, "Are you going to believe the false testimony of commoners against the word of the King's own representative?!"

"I am the King, remember," said King Henry, "and whether you represented me or yourself is the matter we shall fathom. Will you submit to trial by jury, or face the ordeal?"

Now the ordeal to which Henry referred was one of the old laws of the kingdom wherein a man might be pressed by weights until he confessed or expired. Since trial by jury was optional, a man still had the right to refuse it and face the ordeal. Suddenly the Shire-reeve's bravado collapsed, and he sank to his knees begging for mercy.

"You showed no mercy when you drove these people from their lands. Yet you hounded them further in the Shirewood, thinking to end their miserable lives before they should discover the pretext by which you illegally gained control of their lands. In this murderous endeavor you were aided by the Earls of Nottingham and York, and the knights of their shires, who illegally entered with the same murderous intent, and without permission or knowledge of the Crown, those royal hunting estates known as the Shirewood. Therefore, you are hereby ordered by royal decree to appear before the *curia regis* this very afternoon, along with those same earls, whose cases will be heard them."

After dismissing the court of the exchequer, King Henry called Bertie to him and asked, "How is it the subjects of this land are taxed in the future?"

To him Bertie explained the property and income taxes.

"Of course," said the King, "that seems most reasonable. Then if a man own much land which cannot be tilled, or if it lie fallow, so that he hath not the fruits of it, then he should be taxed only on what he hath derived from it!"

And so it was, at the suggestion of none other than Bertie in 1163, that the King would come to levy just such a property tax in 1166 to raise funds for the defense of the Holy Land.

"One question more," said the King to Bertie. "How is the Crown to make a fair assessment of a man's income and property?"

"A jury of his neighborhood peers would know," replied Bertie, "because they would have been dealing with him all year."

"Of course," said the King, "how very simple."

At the end of the day, when King Henry had heard all evidence, he gave his verdict.

"The lands of all the freemen expropriated by the Shire-reeve are to be returned to them. Lands of the Shire-reeve legally owned are forfeit, to be evenly divided among the freemen he wronged."

Marian hugged Bertie, for this meant that once again she would be a well-to-do landowner.

"Now as to the murders of certain people of the Shirewood," continued the King, "for which the Shire-reeve and the Earls of the Shires of Nottingham and York are blameworthy, the penalty for the Shire-reeve is death. For the Earls, the penalty is banishment to France, whence they came, and their lands are forfeit to the Crown."

A wild cheer went up from the King's people. When it had subsided, he continued his verdict.

"Now since the knights of those same Earls, in committing their crimes were ignorant of the true nature of their deeds, believing their combat to be noble against rebels against the Crown, and so ordered by their lords and carried out in knightly servitude, it is my judgment that those who would continue to serve their false lords may follow them in banishment to France, and those who would serve me may keep their knightly fiefs by swearing new allegiance to King and Crown."

Now the knights in attendance cheered King Henry's verdict. In his stratagem he had deprived the banished Earls of their armies, thereby preventing their return to England with armies of rebellion.

"But two more tasks remain this day," said King Henry. "The shire of Nottingham is now without a reeve to do the business of the Crown. In grateful thanks for services rendered in saving the lives of many people in the protection of the Shirewood, I hereby appoint as Shire-reeve Robin of the Wood."

Another great cheer rose to the rafters of the hall.

"Robin is to have the same annual salary as the former Shire-reeve, and as his agent and assistant at one half that same stipend, the Crown appoints Little John. He is to keep Prince Richard's royal pony for his own as a further reward for bringing to the Crown's attention the perilous circumstances of the Shirewood's people when under siege."

At first disbelieving laughter broke out among the people, but Robin strode to the side of Little John, clasped his hand warmly, and the people knew that the King spoke not in jest but truthfully.

"My Lord and King," said Robin, "I beg a word of you."

"Take two, dear Robin," replied the smiling King.

"We thank you as a people for the justice given us, and pray that if this England survive a thousand years more that you shall be known as Henry the Just. In the light of your justice, I would beg one more boon of you."

"Speak," replied the King.

"Some serfs and slaves," said Robin, "who now reside in the Shirewood and have no lands to go to, can yet make do living upon those gifts of deer with which the Crown may choose to present them through myself as Shire-reeve, and also from the well-tended garden, protected from marauding game by a presently illegal system of thickets and hedges. In the interest of the preservation of these people, I ask that you now give the Crown's blessings to those same hedges and thickets."

King Henry thought for a moment and then replied, "Should the royal larder experience a shortage of venison, boar, and other game, as a result of said hedges and thickets, presently illegal, but hereby pronounced as legal, may I count upon you personally, dear Robin, to alleviate that shortage of said game?"

Robin smiled, and the others in the hall laughed outright, for Robin's poaching of the King's game had been well-known. "Indeed, my Lord, you can," replied Robin.

That night there was to be a royal banquet. Bertie, Bongo, and Robin, as well as Little John were invited to attend with Marian. The King had said that there were two more tasks to be completed. Robin's honor was but one, and so all were agog to know the other.

"Don't eat with your fingers!" Bertie commanded Bongo as Bongo tore off a drumstick of roasted quail. Then the King picked up a quail and similarly tore off a drumstick while looking at Bongo. "Okay, eat with your fingers," said Bertie.

At the end of the banquet, when dishes had been cleared, the guests rose to toast the King. Then trumpets sounded and drums rolled. Entering the hall in stately torchlight procession came twelve of the King's most trusted knights. Robin came to Bertie and Bongo, and taking their hands led them before the King.

"Kneel," the King commanded, and kneel they did. Raising his sword, he held it invertedly, making with it the sign of the cross. Then he grasped the hilt and slowly lowered the blade in turn to the shoulders of Bertie and Bongo.

"I dub thee knights of the realm, Sir Bertram and Sir Bongo," intoned the King in high seriousness.

Now Robin still bore the cross-bow, although the deadly crystal bowstrings had been returned to Marian at her behest. Restrung it was superior nevertheless to any other weapon in the world. Robin now presented it to the King.

"It is a gift from Marian and the knights Sirs Bertram and Bongo," said Robin, handing the bow to the King.

"It was a gift to us from Nilrem," replied Marian. "It saved our lives. Where is Nilrem, your Fool?" she asked, having noticed his absence from the very beginning of the banquet.

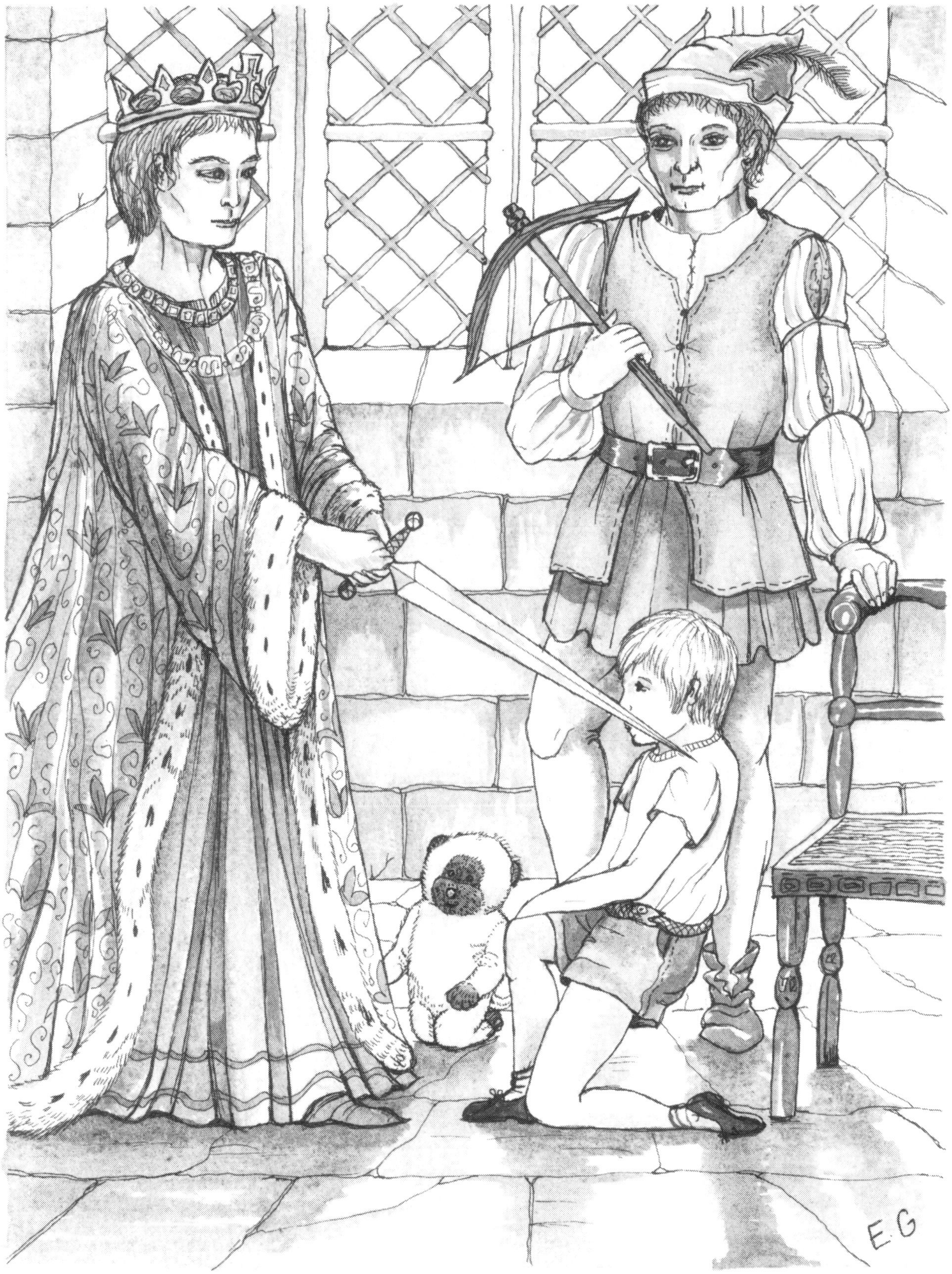
E.G

"Alas," replied Queen Eleanor, "Nilrem expired on the very day you were besieged in the wood. He asked to be buried in a grave next to where Billy Budd's heart lies."

Now Marian wept openly, and so too did Bertie and Bongo, for Nilrem she knew had traded his life for theirs. Hecate had yielded the secret of the bow of death only by taking a life in proportionate sacrifice. Perhaps Marian knew also as did Bertie, that Nilrem was Merlin, "wearing the garb of foolishness," as he had told Bertie upon first coming to him.

That night Bertie and Bongo were given their own chamber, for as knights of the realm it was not mete that they sleep within the same chamber as a female. But Marian kissed them both goodnight, and gave them each candles to light their way down the gloomy corridor to their beds of straw matting, for knights eschewed the comforts of the world in order to toughen themselves for mortal combat.

When they had lain down on the cold straw without blankets in the fireless room where the wind whistled through the open shooting holes in the castle wall, Bertie said to Bongo, "All in all, I think I prefer being not a knight to a knight. But if you think I'm going to call you Sir Bongo, you're crazy."

Bongo patted the top of his head as if to say, "Good knight."

Sometime in the night, Merlin came unto him. Bertie sat up in bed beside his sleeping self. Bongo too sat up, while his sleeping self slumbered on.

Immediately Bertie saw that Merlin's form was changed. Heretofore, he had appeared transparent like a ghost. Now his body shimmered with light, as from a thousand stars. There was a field of energy emanating from him like a dynamo that sent warmth and a feeling of well-being to Bertie, much like what he had experienced in the presence of Pan in *The Wind in the Willows*.

Then Bertie thought sadness at the death of Nilrem.

Merlin's thought came back immediately. "Death is but a veil between lives, into which one moves as easily as passing into a new room.

"The One expands from room to room, life to life, star to star, constellation to universe.

"You are not forsaken or forgotten, nor is the most finite particle of the One.

"All is ever evolving in a widening spiral of higher consciousness.

"I have seen your mother and father and they know you are coming.

"And Bongo's boyhood approaches ever nearer.

"The time has come to end this adventure. Now while it is still dark, go down and out of the castle. I give you the stars of morning as your cosmic inheritance. Follow the Dog Star where he may lead you. *Adieu* until the new dawning of spirit."

When Bertie and Bongo stole down the stairs and passed through the banquet hall it was as if they were in a fairy tale of frozen time. The fire did not flicker on the hearth and the servants all seemed to be asleep, although many stood caught in uncompleted action. Through the outer courtyard they moved, past guards standing like statues of warriors, into an open field where in the sky the Fool's Dog seemed to dance on his hindlegs, beckoning them over a cliff into a new life, or onward to more adventures. Gradually the Dog ceased his dance, and they made a steady march towards the southwest.

   Behind him, the England of 1163, Robin, Marian, King Henry II, and all of his people, the castle, the rocks and trees, took on the two-dimensional form of a tapestry hung on a wall, wavering, shifting, undulating, then departiclizing, taking on the shape of a wave, distorting like the figures in a fun house mirror, finally slowly spinning in an ever-tightening, ever-accelerating gyre that spiralled downward like the grains in an hour glass, until it disappeared into a vortex of nothingness.

CONTINUED IN VOLUME TWO

GLOSSARY
BOOK FOUR

aggregate: a gathering of parts into a unity or mass
agog: highly excited by eagerness
archbishop: next to the pope, the highest church official
atrocity: an extremely cruel or brutal act of violence
axis: the central point around which an object revolves
balmy: producing balm by oozing from the bark or leaves
besieged: attacked
bilious: a greenish-yellow color like the bile coming from the liver
bizarre: strange
black humor: black bile
blaspheme: to speak irreverently of sacred things
boding: predicting as by an omen or sign
bravado: a phony show of courage
canopy: an overhanging covering
cantered: galloped on a horse
carnival: a festival or amusement show
chaff: the husks of grains or grasses
champion: one who defends another or a cause
coarse: lacking refinement or taste in manners
crowned: made king
crystalline: clear like crystal
Cupid: the god of love
curia regius: Latin for the king's councils or administrators
decimated: destroyed in great numbers
deliverance: the act of setting free or liberating
denizens: inhabitants or people living somewhere
dexterity: agility or skill in using the hands or body
diplomatically: tactfully
discreet: safe or cautious
disenfranchised: to deprive of territory or the rights thereof
dissent: disapproval
donned: put on
ecclesiastical: of the church or its clergy
endowed: provided with
ensconced: securely settled on
entrails: the internal parts of a body
eschewed: avoided or shunned
exchequer: a treasury or money collecting department of a government
expropriated: taken possession of
extinguished: put out, as a light or fire
fallow: unplanted or uncultivated
familiar: a witch's animal companion
fathom: to get to the bottom of
fermented: changed into part alcohol
fiefs: territories or properties held in fee
forays: sudden attacks
forester: an officer responsible for the care of a forest
fusillade: a continuous outpouring of anything
galaxies: a large cluster of stars
goaded: prodded or driven as if with a stick
gut: an animal entrail
gyro: a circular course or motion
haggard: looking exhausted
helm: a medieval helmet
idiom: a style of speaking peculiar to certain people
illuminated: lighted

imbues: inspires or impregnates
imputation: an act of accusing someone of a fault or a crime
incarnated: embodied in the flesh
inexorably: unyielding
ingenuity: cleverness with concept or design
initiatory: initial or first
issue: an offspring or child
Januarius: the Latin name for January
jousting field: a place where jousts or combats are held between knights
larder: a place where food is kept
liege: a lord or ruler
litany: a long account or recital
machinations: schemes or plots
maidenhead: virginity
marauding: raiding
meandering: winding
mettle: courage
minions: servant favorites of a lord
Normandy: from the Normandy area of France
numinous: mysterious or supernatural
obverse...proposition: a form of inference in which a negative is obtained from an affirmative, or vice versa
paces: steps used to mark distances
parried: blocked or avoided
phalanx: a closely massed body of troops
phosphorescence: glowing in the dark
pillage: to violently rob
pique: to interest or stir one's curiosity
pith: the essence or essential part
prattle: chatter or foolish talk
prophesied: predicted or told before actually happening
puer: Latin for boy
pun: a funny use of one word that sounds like another
python: a large snake that drops from trees upon its prey
rampart: wall
reflectively: with thought
scowl: to draw down the eyebrows, to look angry
scrutinizing: to view or look at closely
sidled: moved sideways
siren: a mythic creature who lured sailors to destruction by singing
slaughter: the killing of great numbers of animals or people
soothsayer: one who foretells the future
squelched: to squash or crush
stalwart: sturdy or valiant
stratagem: a plan for gaining an advantage
stymied: blocked or prevented from carrying out a purpose
supple: bending easily
surreptitiously: secretly
temper: combine or blend
tempered: imparting strength
thrall: enslaved or in bondage
travail: trouble
try: test
venison: deer meat
villein: serfs, or persons of low birth
visages: appearances or aspects
warp and woof: the interlaced elements or particulars
waylay: to lie in wait for or ambush
zephyr: a gentle wind